A Prehistory of South Florida

A Prehistory of South Florida

WARREN ZEILLER

McFarland & Company, Inc., Publishers
Jefferson, North Carolina, and London

Library of Congress Cataloguing-in-Publication Data

Zeiller, Warren, 1929–
  A prehistory of South Florida / Warren Zeiller.
    p.   cm.
  Includes bibliographical references and index.

  ISBN 0-7864-1971-7 (softcover : 50# alkaline paper)

  1. Geology — Florida. 2. Paleontology — Florida.
  3. Florida — History. I. Title.
  QE99.Z45   2005
  557.59 — dc22
                                                                2004023951

British Library cataloguing data are available

©2005 Warren Zeiller. All rights reserved

No part of this book may be reproduced or transmitted in any form or by any means, electronic or mechanical, including photocopying or recording, or by any information storage and retrieval system, without permission in writing from the publisher.

On the cover — *foreground*: a compacted mass of fossilized organisms, shells and corals confirms the marine origin of the Floridan Peninsula; *background*: the Clovis Point displays a high degree of craftmanship

Manufactured in the United States of America

McFarland & Company, Inc., Publishers
  Box 611, Jefferson, North Carolina 28640
    www.mcfarlandpub.com

Acknowledgments

I wish to express my gratitude to Bob Carr, archaeologist and director, Miami–Dade Historic Preservation Division; Irving Eyster, Society of Professional Archaeologists, Key Largo; Mario Ferrante, Miami Museum of Science & Space Transit Planetarium; Dan Foxen, Everglades National Park; Richard G. Haiduven, archaeological consultant, Miami; John A. Gifford and Marty Healy, Rosenstiel School of Marine and Atmospheric Sciences, University of Miami; James Harold Hudson, regional biologist, Florida Keys National Marine Sanctuary, Key Largo; Jim Lord, Archaeological and Historical Conservancy, Miami; William R. Maples, the C.A. Pound Human Identification Laboratory, and William H. Marquardt, Florida Museum of Natural History, Gainesville; Jim Miller, and Louis D. Tesar, Florida Department of State, Bureau of Archaeological Research, Tallahassee; Lee Ann Newsom, Center for Archaeological Investigations, Southern Illinois University at Carbondale; Sheila Saltis, the Graves Museum of Archaeology and Natural History, Dania; Richard R. Souviron, Coral Gables; Irene Swilp, Stones & Bones, Hilton Head, S.C.; Paul Theerman, Smithsonian Institution, Washington; and Ed Thompson, photographer, Miami. All willingly shared their expertise, experience, data, photographs, and time with me on behalf of this work.

Special thanks are due Miami–Dade County archaeologist John Ricasak, who invited me to work as a volunteer on the fascinating Miami Circle feature at the Brickell Point Site.

Contents

Acknowledgments	v
Preface	1
1. The Floridan Plateau	9
2. The Paleoindian Period: Before 8,000–7,000 B.C.	17
Cutler Fossil Site	30
Monkey Jungle	34
Warm Mineral Spring	38
Little Salt Spring	41
3. The Archaic Period: 7,000–2,000 B.C.	50
Weston Pond	51
Little Salt Spring	51
Bay West Site	54
Horr's Island	56
Atlantis Site	58
Santa Maria Site	59
Cheetum Site	61
Markham Park Site No. 2	61
4. The Formative or Ceramic Period: 2,000 B.C.–A.D. 1513	74
East Okeechobee Cultural Area	
Margate Blount Site	81
Lake Okeechobee Cultural Area	
Fort Center	83
Ortona Site	89
Caloosahatchee Cultural Area	
Pine Island	90
Useppa Island	96

Contents

Josslyn Island	98
Horr's Island	98
Mound Key	99
Ten Thousand Islands Cultural Area	
Big Cypress	104
Key Marco	104
The Everglades Cultural Area	111
Cape Sable	113
Bear Lake Mounds	114
Homestead Site	115
The Florida Keys	117
Biscayne National Park	122
Snapper Creek Site	124
Cutler Burial Mound	126
Indian Creek Site	126
Arch Creek	128
Peace Camp Site	129
Madden's Hammock	130
Granada Site	131
Brickell Point Site	139
Key Biscayne Sites	149
5. The Historic Period: A.D. 1513–Today	**152**
Postscript	178
Glossary	183
Appendix I. An Act to Regulate Trade and Intercourse with the Indian Tribes	189
Appendix II. Treaty with the Florida Tribes of Indians (Moultrie Creek)	192
Appendix III. Treaty with the Seminole (Treaty of Payne's Landing)	198
Appendix IV. Metropolitan Miami–Dade County Historic Preservation Ordinance	201
Appendix V. Federal and State of Florida Statutes Relating to Archaeological Investigations	206
Bibliography	209
Edible Botanicals Bibliography	217
Index	219

Preface

A dozen years ago during a speech I was presenting to the Rotary Club of Homestead, questions arose regarding the prehistory of this rather unique South Florida geographic locale. A silence fell upon the room. The years from man's entry onto the Floridan Peninsula to the nineteenth century were a complete mystery.

Located approximately 30 miles south of Miami, Homestead and neighboring Florida City are the southernmost cities on the continental United States. Immediately west is the main entrance to Everglades National Park, and a couple of miles east is Biscayne National Park, our nation's newest underwater national park. This semitropical agricultural area is best known internationally for tourism to its two national parks, and to other local family-oriented attractions. Farming dominates the land east of the glades, the nation's winter breadbasket. It also is famed for its tropical plant, bromeliad, and spectacular orchid nurseries. To the south lies the beautiful, fascinating, 127 mile arc of the Florida Keys.

Homestead is the point from which, in the late 1800s in a marvel of engineering, Henry Flagler extended his railroad, bridging broad spans of the Atlantic and crossing islands all the way south to land's end in Key West. Apparently, all in the Rotary group, including me, had little more than a rudimentary knowledge of who or what had preceded the indefatigable Flagler.

Since their land is a prolific source of marine fossils that are everywhere at the turn of a harrow, plow, or trowel, the Rotarians' interest ranged from geological formation of their unique Florida peninsula millions of years ago, to the much later source, arrival, distribution, expansion, and lifestyle of the peninsula's earliest people. And well it should; the peninsula was beneath the sea a number of times, so the former marine environment necessitated rather specialized human adaptation. According to Alvah Simon in *North to the Night*, his book about the Arctic people known as Inuit (Eskimoes) who survive in a very different environment: "To truly understand a people, we must first understand the land in which they live — that is, get a hands-on feel for the demands it places on them and the rewards it bestows." Although at the opposite end of the environmental scale, the Floridan Peninsula, especially at its southern extremes, certainly proved an equally challenging environment: extremes of heat, not cold; sand, not snow; highly efficient predators such as huge Pleistocene lions, saber-toothed cats, and dire wolves instead of polar bears, orca, and walrus.

Our curiosity had been piqued, mine in particular. An avocational archaeologist, I would move heaven and plenty of earth to collect, view or touch any prehistoric thing, plant, animal or evidence of past cultures, especially anything made by human hands. To hold such treasures was and still is fascinating, almost an obsession. Then I am driven to learn everything about it, to insert myself into that life, to attempt to accomplish what that person or persons were able to do, with what they had to work with, at that given time in our prehistory.

Naively, I accepted the challenge to research the questions posed by my fellow Rotarians. It was their inquisitive good nature that stimulated my interest in the evolution of the South Florida land and its fascinating earliest people. My hope is to share the wonder of the ancient settings described, the clever adaptations of the people to their changing environment, and the drama of human survival in an often hostile land.

The science of anthropology, inclusive of various facets of its disciplines, would provide many of the answers we sought. Tim Ingold, editor of the *Companion Encyclopedia of Anthropology* (1994), states: "Anthropology, as it exists today, is not a single field, but rather a somewhat contingent and unstable amalgam of subfields, each encumbered with its own history, theoretical agenda and methodological preoccupations." He explains that anthropology is defined as the study of human evolution, with four subfields generally recognized today: physical anthropology, the study of evolution of the human body; cultural and social anthropology, involving evolution of human beliefs and practices in which study of contemporary primitive societies is utilized to provide insight into conditions of modern civilizations; and archaeology, which studies the evolution of material artifacts.

Ingold further states that "a synthesis of our knowledge of the conditions of human life in the world, in all its aspects, is something worth striving for, and working towards such a synthesis is the essence of doing anthropology."

That was a very tall order for the type of research I wanted to do.

Locally available research facilities were the Miami–Dade Public Library system, as well as the University of Miami library, the Miami Museum of Science and Space Transit Planetarium, the Homestead Pioneer Museum, the Historical Museum of Southern Florida, and archaeologists of the Miami–Dade County Preservation Division and the Graves Museum of Archaeology and Natural History in nearby Dania. I could also rely on newsletters of the Archaeological and Historical Conservancy and the Southwest Florida Project, the Institute of Archaeology and Paleoenvironmental Studies, the Florida Museum of Natural History, the University of Florida and the Florida Anthropological Society, both in Gainesville, plus the state's Division of Historical Resources in Tallahassee, and the wealth of data on the World Wide Web. With these resources at hand, I hoped the information I sought would not be too difficult to find. I dug in for what turned out to be a part-time 15-year study.

Prior to the writings of early European visitors to the North American continent, our earliest tangible information on the area's human inhabitants is gleaned from archaeological research and oral traditions of American Indians. The term "archaeology" comes from the Greek word *archaiologia*, meaning archives, antiquarian lore. Unlike the other three subfields of anthropology, archaeology must reconstruct ancient life through recovered artifacts and other evidence of the human past. Archaeologists excavate the earth with extreme care. They must determine material, size, shape, decoration, and myriad other aspects of an uncovered artifact, as well as record in infinite detail exactly where it has been found, how deep, in what relation to other artifacts, etc. These data define the artifact's context. Past human diets and environments can be defined through other collected materials such as animal bones and bits of wood, plants, seeds, and coprolites. Charcoal and other organic materials are preserved for radiocarbon dating. Soil colors are recorded, changes in which might have resulted from human activity such as a garbage pit; small dark circles might indicate habitation postholes. For every hour in the field probably ten or more are involved processing the data in the lab. This study of the material remains of past human life provides windows into lives of antecedents. It also helps us understand the process of culture.

Excavation of an archaeological site does involve its systematic destruction; the original relationships of artifacts and features never can be restored. Therefore, field archaeologists must record precise vertical and horizontal positions of all significant objects (artifacts, ecofacts, etc.) within the site. They are documenting the objects' "context." All future study has to rely on these recorded verbal and numerical descriptions, maps, drawings, and photographs. Based upon that research and recovered remains, such as bones and teeth, tools, pottery, garbage, remnants of shelters or buildings—all those things that constitute the ancients' material culture—archaeologists are able to offer careful inferences regarding the aboriginals' food, how they lived and exploited their environments, and how their cultures evolved through the millennia. Why do people change or stay the same? How and why do societies change from small nomadic bands of related people to highly organized chiefdoms? In brief, archaeologists attempt to develop a picture of the past. How do these findings compare with American Indian oral traditions? Do the ancients' successes and failures provide lessons for us today? To accomplish this task, Renfrew and Bahn (1991) summarize:

> the materials the archaeologist finds tell us nothing directly in themselves. The scientist collects data (evidence), conducts experiments, formulates a hypothesis (a proposition to account for the data), tests the hypothesis against more data, and then in conclusion devises a model (a description that seems best to summarize the pattern observed in the data).

While numerous artifacts are mentioned in this text and many are illustrated, in general the text does not include lengthy written technical descriptions of each artifact, their weights and measures, or information on the quantities of specimens per site. Discussions of archaeological field methods and examinations of anthropological contexts are excluded as well. While extremely important to the scientist, that type of material is less appropriate for a general text. The bibliography will enable a serious student to locate these data with ease.

Archaeologists generally refer to the southern portion of the Florida peninsula's human habitation time frames as cultural periods. They divide and title them as Paleoindian, Archaic, and Formative or Ceramic, the latter in the southern end of the peninsula being further divided into Glades I, Glades II, and Glades III. The time spans within given periods do vary somewhat according to different scholars and there are even time subdivisions within the Glades sequence.

Division of the southern portion of the Florida Peninsula into cultural period geographic boundaries also is subject to variations in individual scientists' interpretations. The boundaries generally are based upon environmental variations and the human adaptations thereto. Archaeologists' persistence with these variations is indicative of the dynamic nature of this interesting scientific discipline. Since the bulk of this work leans toward the more southeasterly portion of the peninsula, the cultural periods, time frames, and geographic boundaries will follow those outlined in 1987 by former Miami–Dade County archaeologist Bob Carr.

As I anticipated, my initial trips to the University of Miami library and other ready sources revealed an overwhelming volume of data. The thousands of archaeological sites known throughout the distal end of the Floridan peninsula are far too many to cover in any given text. While I happily pored over endless volumes of scientific research, obviously there was no way to include it all. So, each site presented has been selected to serve as an example within a given cultural period or, in some cases, stretching through a number of cultural periods, within the given geographic boundaries of the area's prehistory.

Immediately it became clear that one particular point of confusion required clarification. In reference to time, scientists often utilize the initials B.P., meaning "Before Present," for ancient or "deep" time prior to 10,000 years ago. This time is based upon radiocarbon analysis of organic materials; the "present" in the term is the year 1950, when the test was devised.

Radiocarbon years are determined by a technique that measures, with an acceptable degree of accuracy, uncontaminated organic specimens such as charcoal, plant remains, bone, and shell by determination of remaining carbon 14, radioactive carbon, in a specimen. This technique is based on the known decay rate of the relatively rare naturally radioactive carbon isotope with atomic mass 14 and a known half-life of 5,700 years. C14's presence in the atmosphere is nearly constant because cosmic radiation replenishes that which is lost through radioactive decay. Radioactive carbon, C14, as well as non-radioactive carbon is absorbed by living organisms during their life spans. After death the C14 decays at a known rate, so the time since death can be determined by measurement of the proportion of remaining C14. Generally, time determined by this method is written and accepted as a given number of years—plus or minus hundreds to possibly thousands.

However, since it is known that the amount of C14 in the atmosphere is nearly constant, not absolute, radiocarbon years cannot be the same as our more familiar calendar years. The time differential can be quite meaningful. With the history of variations of atmospheric C14 reconstructed, the radiocarbon years can be adjusted to the more familiar calendar years using accelerator mass spectrometry (AMS). The variation can be as high as 15 percent. That does not sound like much. But take, for example, a C14 date of 11,500 years B.P. that with AMS adjusts back to a median of 13,350 cal B.P. (calendar years Before Present), and the differential of almost two millennia is substantial. The more familiar calendar years from deep time to 10,000 years ago (cal B.P.) is more understandable and will be utilized herein.

The standard B.C. (before Christ) and A.D. (anno Domini, meaning, in Latin, "in the year of the Lord"), from 10,000 years ago to the present, also are more familiar time measures and will be followed throughout the text.

Numerous methods of approximate time determination based upon age of artifacts or ecofacts are known: the above mentioned radiocarbon dating and accelerator mass spectrometry, plus tree ring analysis (dendrochronology), thermoluminescence, and electron spin resonance, to cite an additional few. All are tools that enable the archaeologist to develop and verify time sequences within the past. Since time that extends beyond our own personal life span is rather difficult to imagine, and a single generation of human life is an infinitesimal fraction of nature's overall scheme of things, time comparisons with human developments and events in other parts of the world are inserted as footnotes throughout the text. For example, about 4,000 years ago when Florida fishers, hunters and gatherers first were struggling with clay, fiber tempering, and ceramic firing techniques, people of the Old World long had mastered the potter's art some 8,000 years before. Egyptians and Sumerians were practicing agriculture, animal husbandry, and using wheeled vehicles; earliest Floridians of that time had not the slightest inkling of those things. The examples are presented to draw a comparison between the Old World state of human advancement and that of the more primitive Florida aboriginal at any given time.

Of the thousands of archaeological sites known to exist throughout the state, many have been destroyed by natural forces, such as erosion. Man certainly has destroyed more than his share. Deep solution holes have been filled in for a variety of reasons, more than likely entombing a wealth of archaeological treasures within. Many Indian mounds have been dug up and hauled away, or plowed and planted since the lime of the disintegrating shells, slowly decomposing bones, charcoal from ancient fires, and humus from past vegetation are ideal foods for many growing plants. Endless cubic yards of shell have been removed from ancient mounds for modern road building. Developers, avocational pot hunters, and vandals, too, continue to take their toll. On the relatively flat plateau of South Florida, the picturesque home on top of an unusual knoll more than likely owes its elevation to an ancient shell mound or midden underneath. Or, downtown, shoppers may be perfectly innocent of the fact that beneath their feet lies an ancient site containing artifacts that tell the story of thousands of years of human cultural development, sealed in perpetuity under the macadam of a parking lot. People must realize that once thoughtlessly destroyed, these bits of our heritage are lost forever.

Although a study of prehistory, this work must touch upon the Historic Period in an effort to discover the source of modern tribes living within the area. Are they direct descendants of the ancient peoples who inhabited South Florida for thousands and thousands of years? If not, who are they and what happened to their antecedents? History of the modern tribes often notes that they are the only ones without treaty with the United States. Study reveals several treaties, most often referred to only by colloquial name, that were broken and thus voided. They are presented in full text within the appendices for interested readers. Federal, state, and local laws governing archaeological investigations are included in the appendices as well.

Material presented herein should prove more than adequate to answer many questions and, possibly, stimulate further research. I hope it will be of some assistance to those interested in site preservation, and prove sufficient to satisfy the many questions eagerly posed by scholars as well as my fellow Rotarians.

1

The Floridan Plateau

The theory of plate tectonics hypothesizes that great sheets of rock, many miles thick and thousands of miles across, carry continents as they slide across earth's semi-molten interior mantle. This movement is referred to as continental drift. Hundreds of millions of years ago, continental drift separated a giant landmass, which scientists call Pangaea, into the African and North American continents. The space between the continents became the Atlantic Ocean.

The Floridan Plateau (the peninsula referred to today as Florida) remained with the North American continent. Eventually, its emergent land resembled a broad sandbar comprised of chemicals and minerals, as well as shells and bones of marine creatures. Deposited over millennia by ocean currents, these precipitates alternately dried and flooded during glacial intervals, creating hardened, overlapping layers of highly porous limestone and dolomite rock. Successive glacial inundations substantially altered the shape of the plateau.

Some 50 million years ago, the peninsula was only an island south of North America. By 10 million years ago, that island again was beneath the surface of the sea. The emergent-flooding cycles increased in frequency between 12,000 and 10,000 years ago, during the latter millennia of what is referred to as the Ice Age. Each successive cold period, lasting thousands of years, froze immense volumes of seawater into an unimaginable ice mass, drawing down sea level and exposing more and more land.

During the ensuing millennia of emergence, rains flushed sea salt from the land and lakes. Existing life forms were able to evolve and expand on the larger land mass. As the climate warmed and the ice masses melted, releasing water back to lakes, rivers, streams, and the rising sea, life forms adapted, retreated, or became extinct. The geological record clearly displays that each successive flooding covered less of the land mass than the last. The glacial-interglacial cycle repeated itself four major times, leaving

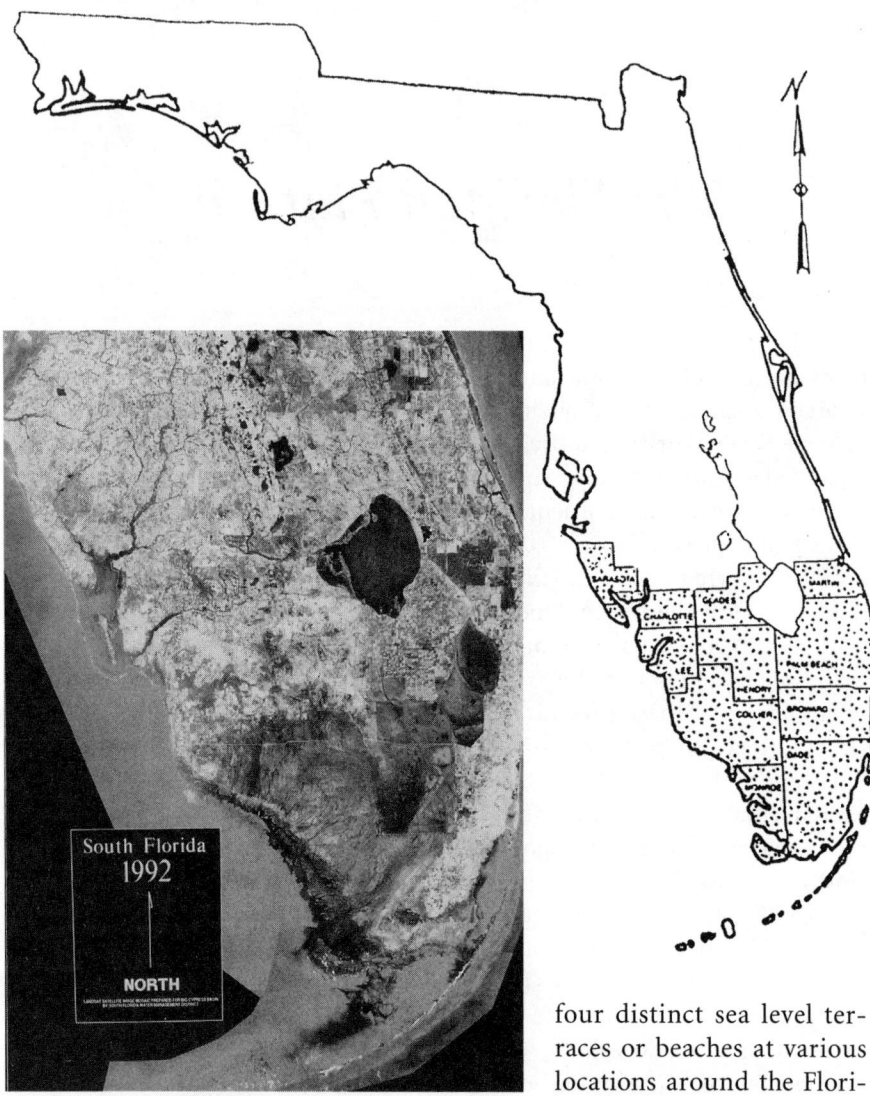

The southern tip of the Floridan Peninsula, the southernmost part of the continental North American landmass, is displayed in the sketch and Landsat composite photograph above. Lake Okeechobee dominates the center of the mass. The dark area south of the lake clearly defines today's Everglades. (Landsat poster courtesy of the South Florida Water Management District, West Palm Beach. Not to scale.)

four distinct sea level terraces or beaches at various locations around the Floridan peninsula, and an unknown number of minor times that are not so easily identified.

Evolving biota expanded from the North American land mass south, as well as from South and Central

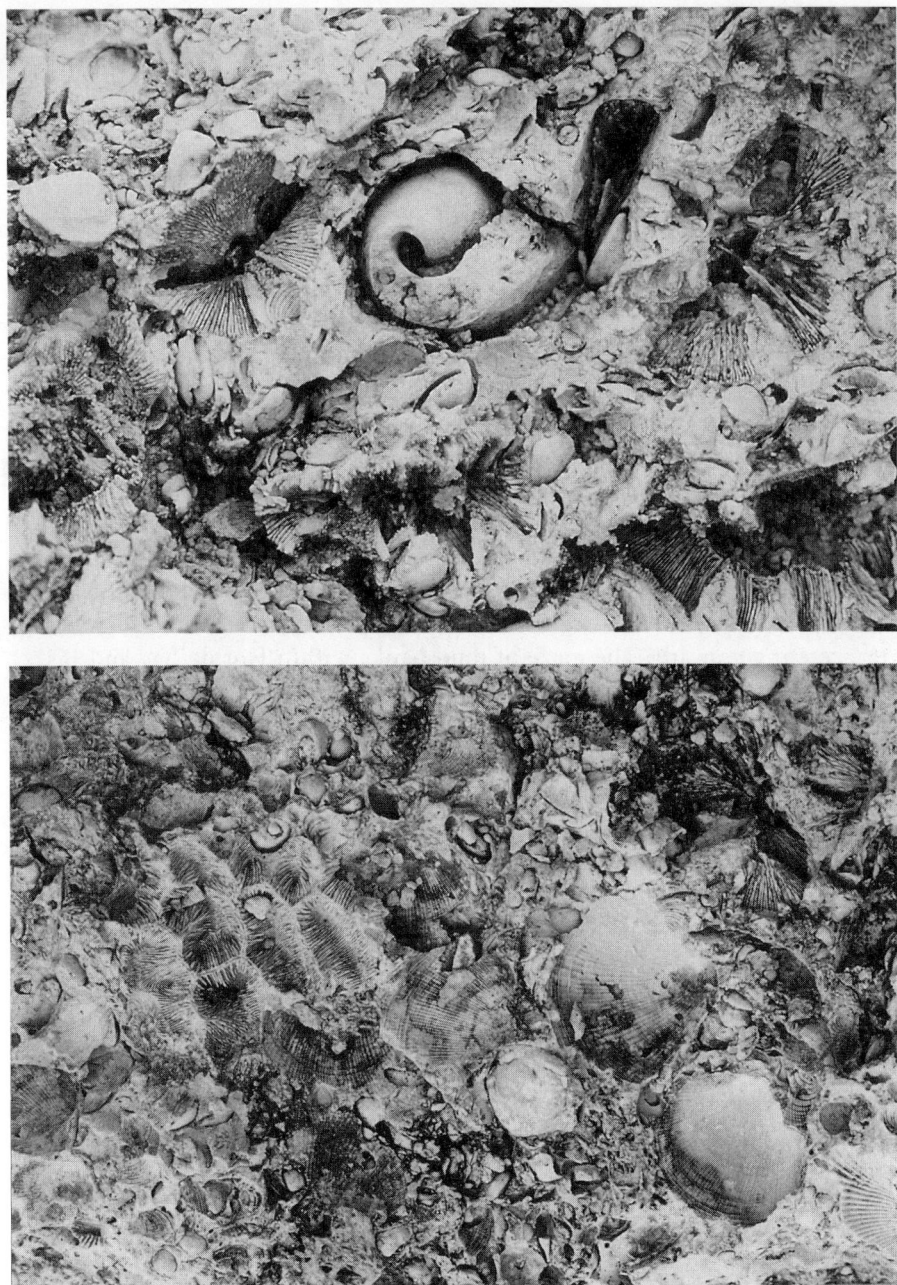

A solid compacted mass of fossilized organisms, shells, and corals imbedded in the oolite matrix confirms the marine origin of the Floridan Peninsula.

America north across southern coasts, and east to the Floridan Plateau. About 6,000 years ago the climate again warmed and Florida began to bloom with flora and fauna commonly found today.

Between 5,500 and 3,200 years ago, the sea level rose from about minus 19½ feet to minus 3¾ feet below current sea level in South Florida. Land mass depressions, such as the Everglades and shallow coastal bays, became more extensively flooded. For example, Miami's Biscayne Bay was little more than a narrow sea channel prior to 5,500 years ago, and was nearly completely flooded by 3,200 years ago. The limestone ridge defining the ocean's seaward margin disappeared beneath the sea, leaving only Elliot Key as a significant emergent barrier. Since that time, relative sea level rise from the last great thawing ice mass (still in progress today, with increasing rapidity in recent years) has averaged only about an inch every 100 years (with the exception of the unusually sharp rise of about 8½ inches recorded for the last hundred, the twentieth century).

This relatively slow rate permits coastlines to stabilize, or even expand through growth of shallow marine environments upward to sea level. The marl levees forming the northern margin of Florida Bay (in the southwest area of Everglades National Park) are largely stabilized and have been accreting seaward. The maze of mud banks within Florida Bay and Biscayne Bay have mostly built up to sea level and set the stage for mangrove colonization. Barrier and patch reefs are catching up to sea level. In some areas, newly flooded inlets contribute to fresh water exchange with coastal bays that eventually causes near-shore reef demise.

Along the southwest coast of Florida, in the Ten Thousand Islands area, the mainland mangrove shoreline stabilized about 3,200 years ago and has since grown upwards. In addition, shallow near-shore environments have become colonized by oyster banks, and these in turn by mangroves.

Along the mainland margin of Biscayne Bay on the southeast coast, some portions of the mangrove shoreline stabilized. Other portions have retreated, probably due to exposure to winter storm winds and waves from the northeast, hurricanes, and the unusually sharp sea level rise. Offshore barrier islands are known to have altered shape, even divided, due to these same forces.

Knowledge of coastline fluctuations makes us realize that what we see today as the state of Florida is only the upper part of a far more substantial barrier that slopes down to the ocean's 50 fathom (300 foot) depth line that marks the edge of a relatively shallow continental shelf.

The Florida Geological Survey designates Miami–Dade County as the southern zone of the Coastal Lowlands. The rock underlying the bulk of this county from the surface down to 30 feet at Miami, to 60 feet about 30 miles farther south in Homestead, is called Miami Limestone. It is composed of macroscopic spheres of calcium carbonate deposited in concentric layers around a core of hard material, such as quartz grain or bit of shell. The spheres are known as "ooids" due to their resemblance to fish eggs. The tiny one-fourth to one-half millimeter diameter ooids are finished as if nature had been creating countless numbers of almost perfect, tiny pearls. Masses of ooids drift to the ocean floor, creating immense dunes within which are vast numbers of shells. Upon emergence from the sea, these dunes consolidate to form multilayered rock known as "oolite." Where vertical oolite surfaces are visible, variations in layering clearly reflect changes in sea currents and other turbulence.

The southern zone of the Coastal Lowlands is naturally divided into two main topographic regions. One is known as the Atlantic Coastal Ridge; it borders the southeastern Atlantic shore as a narrow rock elevation running from the Georgia border all the way south to Homestead. The Coconut Grove section of the ridge immediately south of the city of Miami boasts the highest ridge elevation at about 20 feet above sea level. The ridge slopes abruptly eastward to the Atlantic, and westward more gently to the Everglades, including the Big Cypress Swamp which is the second topographic region.

The low, broad expanse of the Glades is about ten feet above sea level in the northern part of Miami–Dade County. Then it drops almost imperceptibly southward to sea level in a jagged shoreline fringed with thousands of islands.

The limestone rock of the Everglades does contain ooids, but also contains large numbers of massive colonies of bryozoans. These tiny marine invertebrates secrete a calcareous cell in which to live. They reproduce by asexual budding to form knobby colonies a foot or more in diameter. They are encrusting creatures, so the colonies vary tremendously in shape depending on the character and form of the encrusted organisms. When sectioned, the knobs and branches often are hollow, the organic material on which they encrusted having decomposed, leaving only the tubes of multi-laminate bryozoan skeletons. Calcareous worm tubes also are numerous, as are pellets, tiny elliptical calcareous grains believed to have been excreted by marine worms.

The bryozoans (of organic origin) underlie the ooids (of inorganic origin) of the Atlantic Ridge. Together they comprise Miami Limestone.

Current aerial photographs of the Great Bahamas Bank clearly display massive underwater dunes. It is apparent that the foregoing description of the formation of oolitic and bryozoan limestone at the southernmost end

Top: The Atlantic Ridge is clearly displayed at Silver Bluff along South Bayshore Drive, Miami. *Bottom:* Tiny ooids that form the ridge seem little more than dust in the author's palm.

of the Floridan Peninsula is repeating itself in the Bahamas and other Caribbean islands today.

The Atlantic Coastal Ridge south of Miami is cut at right angles to its axis by shallow valleys, known locally as glades. Most are only a few feet deep, are flat-bottomed, and are filled with deposits of limey marl four to five feet thick. Before man altered the area with drainage canals, early settlers reported that it was natural for the lowland between the Atlantic Ridge and the pine and rock land from Cutler south to Florida City to be underwater all summer. Today, flood waters most often are carried off through manmade drainage canals; the rich deposits left by the old cyclical flooding are called Perrine Marl. They provide the fertile fields for production of winter vegetables.

A very cold, oceanic, Labrador Current sweeps southward along the Atlantic coast. At the lower end of Florida, it finally disappears under the warm, north-flowing Gulf Stream. Masses of quartz grains and other material eroded from the Appalachian Mountains are carried seaward in streams and rivers to the Atlantic, then are transported south by the Labrador Current. Always opposed by the force of the Gulf Stream, much of the material is mixed and deposited with oolite grains along the oceanic margin of the Atlantic Ridge. These deposits emerge as sedimentary barrier islands. Off the southeastern shore of the Florida land mass, the islands today are known as Miami Beach, Fisher Island, Virginia Key, and Key Biscayne. Upon emerging from receding seas, these deposits became fairly stable and, prior to development, were covered with dense wild plant growth that was stimulated by very heavy annual rainfall. The heavy moisture is created by the conflict between the contending warm Gulf Stream and cold Labrador Current and falls as rain. The water body virtually surrounded by this land configuration of the mainland Atlantic Coastal Ridge and the sedimentary barrier islands is known as Biscayne Bay.

Arcing southwest of the area is another chain of islands formed by emergent ancient coral barrier reefs, known colloquially as Keys. These narrow, small, Upper Keys from Soldier Key south through Marathon have been built up over millions of years by living coral organisms; hence they differ completely from the sedimentary barrier islands.

South of Marathon, from Big Pine Key to Key West, the islands are larger, more irregular in shape, oriented at right angles to the arc, and owe their geological features to their origin as an emerged oolitic bank of Miami Limestone.

Between the thin veneer of Miami Limestone and underlying dense bedrock is a porous limestone rock strata in excess of 18,000 feet thick, within which are immense underground freshwater reservoirs. The largest reservoir underlying Florida and portions of neighboring states is the Floridan Aquifer. Where the covering soil veneer is porous, water easily percolates into the aquifer. When the water is trapped underground between impermeable layers, it boils to the surface through clefts in the rock as natural artesian springs, the volume of which ranges as high as billions of gallons a day. The water is extremely clear and of constant temperature. These incredible springs always have attracted wildlife and, eventually, humans.

The foregoing brief on formation of the Floridan Peninsula is of importance because there is a critical relationship between the environment and human culture. Geology, topography, and climate are factors controlling water, vegetation, and animal life — all the major resources of human society.

This tip of the Floridan Peninsula, the southernmost part of the continental North American land mass, and its earliest people comprise the focus of this work.

2

The Paleoindian Period: Before 8,000–7,000 B.C.

During the last half of the twentieth century, the variety of hominid species represented by the number of fossils discovered worldwide increased enormously. Hominid lineage, of which modern man (*Homo sapiens*) is the only extant species, appears more and more complex with just about each new paleoanthropological discovery. Let it suffice to list here among ancestral hominids the *Australopithecus* species that first appeared on the African continent possibly 4 or more million years ago, and appears to have reached the end of its lineage about 1.5 million years ago. This coincides with a key transition in the evolutionary chronicle, bipedalism, the development of walking upright on two legs. A second major transition was development of tool making; that appears to have begun about 2.5 million years ago.

Homo habilis fossils date from almost 2 million to 1.5 million years ago. That correlates with a dramatic growth of the brain that evolved somewhere between 2 million and 1 million years ago. There also is evidence that within that time, possibly earlier, hominids first began to emerge from Africa. *Homo erectus* ranged from about 1,700,000 to possibly 200,000 years ago.

Homo neanderthalensis, Neanderthal Man, from 200,000 to 30,000 years ago ranged from Africa through the mid-latitudes of Asia and Europe. *Homo sapiens*, anatomically modern humans, 100,000 years ago began to range throughout Eurasia. They probably were fully established in Europe by about 32,000 years ago. Their migrations, like those of just about all biota, are believed to have been related to environmental changes. Their advantage of increased brain power enabled them to develop abstract thought, art, language, music, and other skills.

Another hominid genus, as well as a number of other hominid species

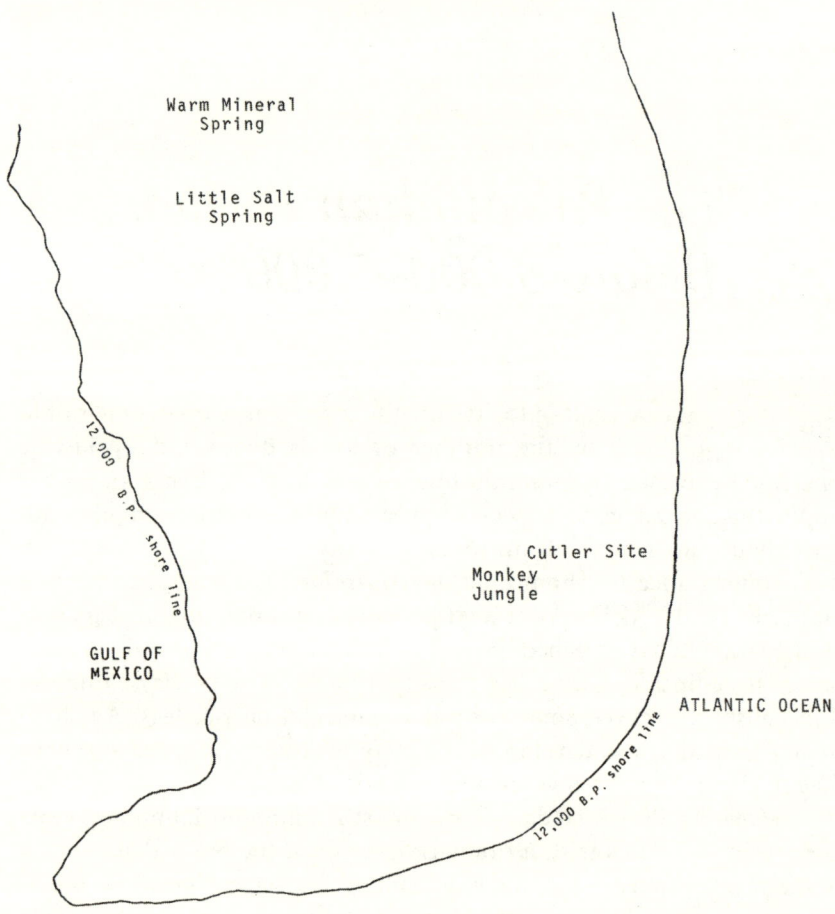

The Floridian Peninsula of 10,000–7,000 B.C. The shoreline depth is at approximately minus 90 feet from that of today. Note the absence of major features such as Tampa Bay, Florida Bay, Biscayne Bay, and Lake Okeechobee. Most recorded human habitation sites are well inland; those that existed nearer the coast are underwater today. (Not to scale.)

within the two genera listed above, tentatively are interspersed among the foregoing, but their designation remains subject to debate. Evidence to date seems to indicate that modern man is just one more branch on a very complex bush, as opposed to the formerly envisioned, more rigid, rather linear "family tree." The prehistory of man has proven to be one of repeated trials, with each evolved species eventually becoming extinct, excepting our own.

Since hominid species have been around for only 4 million or so years, archaeologists will not find artifacts of North American Paleoindians, or for that matter any other humans, in association with fossils of dinosaurs. The latter roamed the earth for as long as 120 million years, then became extinct about 80 million years ago. Cohabitation of dinosaurs and humans happens only in cartoons and B grade movies.

Nor to date have remains of extinct forms of humans been found in the New World, and the likelihood of uncovering such evidence does seem remote. Archaeological data suggest that all fundamental biological and cultural events in human evolution occurred in the Old World, and that by about 30,000 years ago the races were physically similar to humans living today.

The Pleistocene geological period encompassed a time frame of from 2 million to 10,000 years ago, popularly known as the Ice Age. It was during the latter part of this period that the first modern humans, *Homo sapiens*, ventured on to the North American continent, possibly a mere 15,000 years ago. North American incursion of more than 15,000 years ago still is a point in contention, but evidence in that direction may be mounting with some believing that people were crossing to the continent as early as 27,000 years ago.

People are believed to have migrated across the Bering Strait land bridge, known as Beringia. With water locked in the thousands-of-feet-thick Ice Age mass, the bridge became exposed, forming an avenue as wide as 1,300 miles, but spanning only 40 miles in length from Siberia to Alaska. Later when water was released from the melting ice mass, sea level rose, the land bridge was inundated, and eventually disappeared beneath the icy Bering Sea.

During the very latter part of the last Ice Age, from about 70,000 years ago until 10,000 years ago, geologists believe that the western Cordilleran Ice Sheet and eastern Laurentide Ice Sheet drew apart, creating an ice free corridor through which humans eventually were able to migrate. Based upon geological and archaeological investigations of 47 U.S. sites over the past 50 years, several Canadian geologists believe a narrow corridor opened up about 15,350 years ago. By about 14,000 to 13,000 years ago no physical human barriers in the deglaciated areas would have remained. Climatic conditions similar to those prevailing today began to occur, and people came in a number of waves to this New World,

Generally, it has been accepted that the first people came from northeast Asia and Siberia between about 15,000 and 11,500 calendar years ago. Bluefish Caves and Old Crow Basin are archaeological sites located in an Arctic region that would have been part of ancient Beringia. Both contain

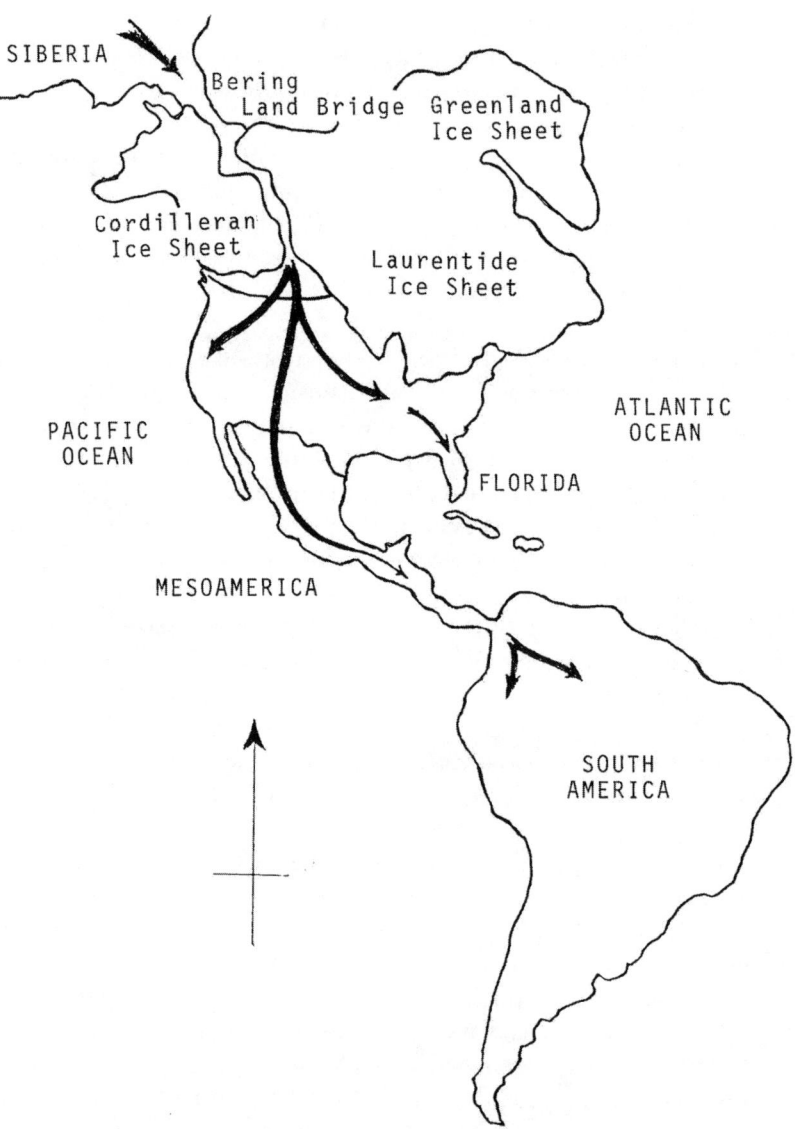

Human migration routes trace their way from Siberia across the ice bridge known as Beringia south through ice–free corridors into North and South America. Possible additional circum–Atlantic and Pacific routes await confirmation. (Not to scale.)

evidence of human presence recently dated to about 25,000 years ago. As previously noted, these data and other site dates determined by numerous advances in that field bring the 15,000-years-ago date into question. Some scientists question whether migrations might have occurred much earlier, possibly avoiding the impassable glaciers by traveling by boat along the Pacific coast. The question remains unanswered. In any case, it is apparent that on reaching the North American continent, some people followed the retreating glaciers north and east, while others made their way south into Meso and South America. Based upon the current archaeological record, they reached Clovis (New Mexico), Brazil, Chile, and Argentina by 11,000 or possibly more, years ago and the southern end of that continent at the Strait of Magellan soon after 9,000, or again possibly more, years ago. Here too, modern dating techniques have brought the time frames into question, suggesting much earlier arrivals by thousands of years.

A second, later human wave is thought to have migrated across Beringia into Canada, then just 600 or 700 years ago moved south to be ancestral to the southwest's Navajo and Apache tribes. An interesting method of tracking migrations is one used by virologists. They find that a virus which lives in the kidneys, called the JC virus, is a strain carried by modern Navajos that is nearly identical to a virus carried by modern day residents of Tokyo. This suggests a close relationship between the two groups, and provides additional confirmation with DNA and human skeletal aspects of anthropological studies.

A third migration from northeast Asia of groups ancestral to modern Eskimos and Aleuts occurred perhaps 10,000 or 9,000 years ago. The archaeological record indicates that by about 10,000 years ago, some of these people had spread across the entire North American continent to the Atlantic coast, even to the southernmost tip of the Floridan Peninsula.

Here is a problem with the generally accepted paradigm. In the eastern part of the continent, Pennsylvania's Meadowcroft Rock Shelter (declared a Commonwealth Treasure by the Pennsylvania Historical and Museum Commission) has been dated back to 16,200 years ago at its deepest level of human occupation. The Saltville site in Virginia and Topper in North Carolina are thought to be of similar age and are under investigation. University of Florida paleontologist David Webb and underwater archaeologist for the Florida Bureau of Archaeological Research James Dunbar reported on a site 30 feet below water level of the Aucilla River east of Tallahassee. The site "yielded thousands of items, including the most extensive collection of worked ivory tools in North America." A 7.5 foot mastodon tusk bore six cut marks at the jaw line where people cut it out of the skull. Dating proved the butchering took place 12,200 years ago, possibly the oldest known animal butchering site on the continent, and 700 years prior to the Clovis sites. Wild gourd seeds (Curcubita pepo) were found in mastodon dung dating back to that time at the site as well. Prior to the find it was thought that the gourds had been introduced by people migrating up from Mesoamerica 7,000 years ago. If these so-called "pre–Clovis" dates are substantiated, Clovis being long recognized as the earliest site in the continental U.S., the occupation of the American Continent timeline will have to be adjusted accordingly.

Some scientists theorize about earlier European incursion onto the Atlantic coast to explain the earlier eastern site dates. In 1999, a controversial new approach to prehistory of the continent was presented by three prominent archaeologists who say that North America's first inhabitants, Solutreans, may have crossed the Atlantic Ocean from Europe's Iberian Peninsula, the area that now is Spain, Portugal, and southwestern France. They do concede that other paleo-explorers reached the Western Hemisphere too. But, they say, the Solutrean people quite possibly sailed to the New World in boats, following the northeastern route along the 1,400 mile fringe of the Atlantic ice sheets that existed 16,000 years ago. And why not; by this time South Seas islanders had been sailing open Pacific waters for millennia. They are thought to have reached Australia across the open ocean more than 50,000 years ago. Atlantic crossings by other people do seem feasible. The researchers base their hypothesis on comparison of the Solutrean's distinctive style of projectile points and other artifacts with those that are virtually indistinguishable from Clovis artifacts. Both used bi-facial pressure flaking technology, spurred endscrapers, bone foreshafts, left caches of points, and etched geometric designs on stones. The researchers state that there is nothing in the Clovis culture that is not found in Solutrea. Their main problem is the time difference between the Solutrean culture of 23,000 to 18,000 years ago and that of Clovis occupation of about 6,000 years later. Where were they in the interim?

Tracing DNA male Y chromosomes, geneticist Spencer Wells has chronicled the path of human migration from India across 150 miles of the Indian Ocean to Australia. This certainly makes possible passage along the Asian Pacific coast, across the narrow Beringia, possibly south along the American continental Pacific coast. Whether humans 15 or 20,000 or more years ago were capable of navigating along perilous frozen Atlantic coasts is another matter. If the earlier dates of Meadowcroft and the other pre–Clovis sites are substantiated, the time differential will diminish. This research presents such a radical new view that it probably will take years to evaluate adequately. Eventually, conclusions may be reached based upon rigorous archaeological and geological criteria established in the early years of this century by Holmes and Hrdlicka, physical anthropologists at the United States National Museum, that require:

> a clearly defined stratigraphy, and a clear understanding of the stratigraphic context of the finds, and the formation of the layers in which they are found.
>
> Reliable and consistent radiocarbon dates, or dating established by some other widely accepted chronological method.
>
> If possible, field and laboratory evidence from other disciplines to support the chronological and geological context. A good example of such evidence would be concordant pollen samples.
>
> The presence of humanly made artifacts in a primary stratigraphic context, artifacts that are established as being of human manufacture according to strict scientific criteria.

Today, a fifth recommendation has been added that states the site should be examined by a team of specialists, including archaeologists, geoarchaeologists, and dating experts. The visit by the specialists should be made at time of excavation when the relationship between stratigraphy and artifacts can be clearly seen.

As Keith Kintigh, president of the Society for American Archaeology, says, "The interplay of theory and evidence involved in achieving a scientific understanding of the timing and nature of the initial population of the Americas is a classic example of the scientific process." How interesting it will be if these most recent pre–Clovis hypotheses withstand critical review.

In any event, the migrants who made it to this new land were intelligent people who brought with them the variety of skills of many, many thousands of years of development, and ecological and environmental adaptation.

Culturally, the earliest Native American people utilized a variety of specialized, well-made stone, bone, antler, horn, shell, and wood implements. They subsisted by gathering botanicals, and hunting a broad spectrum of animals, many of which now are extinct. The land bridge they or their antecedents had crossed had evolved into tundra, marsh, and grassland rich with herds of mammoths, reindeer, musk oxen, horses, bison, and other animals. This wealth of game and botanicals apparently drew them on. Their way is traced by a trail of preserved tools, as well as those formed by striking and flaking stone to a sharp edge (knapping), such as pebble choppers, awls, gravers, scrapers, points, and cutting tools.

Every society, whether prehistoric or historic, is characterized by distinctive material products of its culture. Each product can be classified by its design, its material, and the function for which it was formed. That product can vary in style and attributes, and is termed an artifact. The word "type" refers to an artifact in which several attributes combine, or cluster, with sufficient frequency or in such distinctive ways that the archeologist can define or label the artifact, and can recognize it when she or he sees another example.

Of these implements, the previously mentioned, distinctively fluted, lanceolate-shaped stone projectile or ceremonial Clovis point does serve as an excellent example. It is quite specific in both beauty of form and excellence of workmanship. While its place of origin to date remains undetermined, it was named for the site in Clovis, New Mexico, where the first points of this type were found in this hemisphere. Not found to date anywhere in the Old World (barring substantiation of the Solutrean theory), apparently it was an American development, the trail of which across the continent traces the migration of bands of early humans from Alaska,

The Clovis Point is named for the site in Clovis, New Mexico, where it was found in association with bones of a mammoth. The three and a quarter inch long example displays the high degree of craftmanship, base fluting, and general shape that are features common to all others of this type.

through ice-free corridors to British Columbia, south to northern Mexico, from Nova Scotia to Florida. The Clovis culture was thought to have dated back to 11,500 C14 years ago, but here, too, recent re-testing with modern AMS technology indicates that 13,500 calendar years ago (cal B.P.) would be accurate.

The unique points found in the west knapped by the so-called Clovis People formerly were thought to have been dated earlier than Clovis points uncovered in the east. On the Floridan Peninsula they seem to be concentrated in the northwest area. Farther south they are found rarely in the vicinity of deep sinkholes or springs that penetrate through surface sediments into the ancient limestone. Only recently are there reports of others being found miles offshore beneath the surface of the Gulf of Mexico. Thus, scientific evidence of Paleoindian presence and migration all the way across North America appears to be overwhelming.

The origin myth of a tribe of Native Americans known as the Lower Creeks, passed down by word of mouth (oral tradition), is of particular interest as it details an identical migratory path. Swanton (1922) relates their myth, saying that their ancestors were called the A'teik-ha'ta. Anciently, they were said

> to come to some place where the sea was narrow and frozen over. Crossing upon the ice they traveled from place to place toward the east until they reached the Atlantic Ocean. They traveled to see from where the sun came.... Then the Muskogees, who claim to have emerged from the navel of the earth somewhere out west near the Rocky Mountains, came to the place where the Hitchiti were living. The Muskogee were very warlike, and the Hitchiti concluded it would be best to make friends with them and become a part of them. Ever since they have been together as one people. Hitchiti is the Muskogee word meaning "to see," and was given to them because they went to see from whence the sun came. So their name was changed from A'teik-ha'ta. The two people became allied somewhere in Florida.

Paleoindian society, assuming it paralleled modern bands of primitive peoples, probably consisted of small groups, generally of less than 100 people related by kinship. There probably was little disparity in status among members, with the band lacking formal leaders. They lived on the grassy and swampy plains, and gained much of their livelihood by hunting small animals. Whenever possible, they were able to take stronger and swifter prey through ingenuity and coordinated effort. Such large mammals as mammoths, camels, horses, and an archaic form of long-horned bison, among others, fell beneath their primitive weapons. Fossil remains of such creatures with appropriate stone points lodged in their bones record such successful hunts.

Mastodons and mammoths shared the Floridan Peninsula with Paleoindians. *Top left:* Mastodons stood about eight to ten feet at the shoulder and weighed up to about six tons. *Top right:* Mammoths reached ten to 12 feet at the shoulder and weighed up to eight tons. *Bottom left:* Mammoth's molars show that they were more adapted to grazing savannas and grasslands. *Bottom right:* Mastodon teeth show they browsed the forests and woodlands.

The bands did move seasonally to exploit wild food resources, but the archaeological record of the people's utilization of vegetable materials is sparse; currently archaeobotanists are making rapid advances in this field. Perhaps as well as seasonally gathering edible botanicals, the Paleoinditan bands—like many animals and their human counterparts thousands of years later—consumed the stomach contents of their herbivorous prey. This was an energy efficient way of gaining necessary nutrients and fiber in the diet, especially during times when food otherwise was in short supply.

As Ice Age oceans shrank and the climate became colder, there was

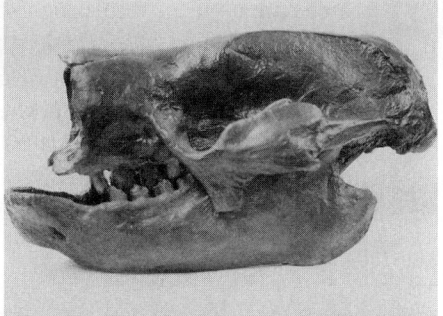
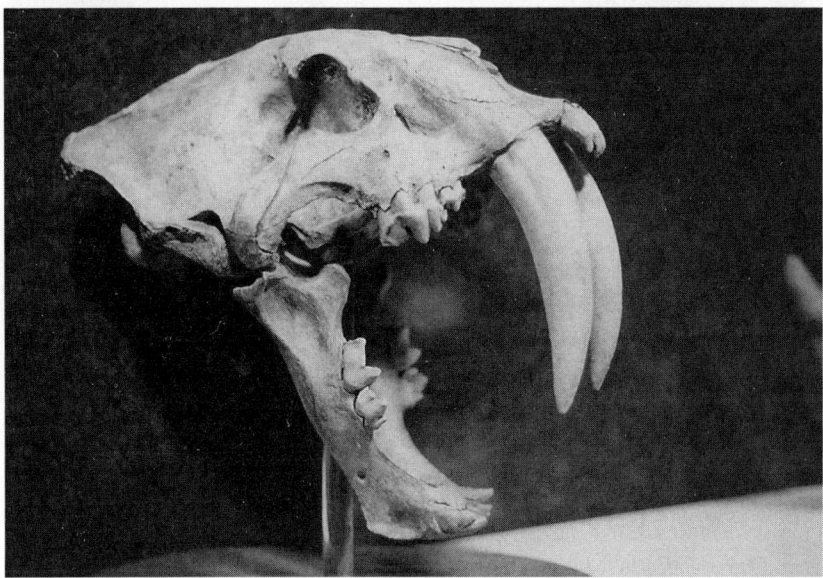

Top left: The American lion. *Top right:* giant round sloth. *Bottom:* Saber-toothed cat. These are but three of the dangerous creatures with which Paleoindians had to contend.

less moisture evaporation, and therefore less rain. Consequently, the continental land mass became more and more arid. Even during the coldest period of glaciation, the two mile thick ice sheet covered only the northern half of the North American continent as far south as Kentucky. The Floridan Plateau remained comparatively warm by comparison to areas under glaciation, and so became a haven for creatures from the north, as well as those from the west and tropical animals from South America. The latter included giant ground sloths and armadillos, capybaras, tapirs, llamas, and peccaries, all of which had migrated south and, due to the warmed climate, were returning to their former range. Carnivores included the hog-nosed skunk, spectacled bear, margays, ocelots, jaguars, saber-toothed cats, Scimitar cats, dire wolves, and hyenoid dogs.

The mammoths, saber-toothed cats, and other mega-fauna all disappeared from earth between 11,000 and 8,000 years ago; the long horned bison species of that time survived perhaps 1,000 years longer, then it too became extinct. The early horse disappeared with the rest, and the modern horse was not reintroduced to the continent until its 16th-century arrival with the Spaniards.

Animals that existed then and survived into modern times include the black bear, gray wolf, red wolf, bobcat, panther, and the bison more familiarly known as the Great Plains buffalo. Florida's gray and red wolves had survived with the rest, but were exterminated by man by the first decade of the 20th century.

The first people on the Floridan Peninsula certainly were not neophytes in a new land. To the contrary, they carried with them the millennia of experience and environmental adaptability of their ancestors. Whatever accomplishments were credited to humans in the Old World — the several thousand years it took multiple generations of them to reach across the new continent — the nomad Floridians brought with them. Their basic need was to develop lifestyle variations demanded by the new and changing environments. Currently, archaeologists surmise that small bands of Paleoindians must have migrated onto the Floridan Peninsula from the north as a result of the natural expansion of human populations, movement of the hunt, curiosity, and the nomadic nature of the Paleoindian bands, to say nothing of their understandably wanting to escape the icy grip on the northern half of the continent.

During the last stages of the Pleistocene Ice Age, Florida's environment was quite different from that of today. The extreme cold in the north that locked vast amounts of water in the ice cap lowered the level of the sea perhaps as much as 300 feet during the peaks of glaciation. Cooler but not freezing temperatures and semi-arid conditions prevailed over the southern Florida Peninsula, a land mass double of what it is today, with the Gulf of Mexico shoreline a good 100 miles west of its current location. The greatly expanded Gulf of Mexico coastal plain, known to paleogeographers as the Gulf Coastal Corridor, was a major migration route for Pleistocene animals. Within their gut they carried, and eventually eliminated, seeds of tropical plants they had consumed. Birds, flying mammals, insects, even the wind provided additional modes of dispersal, if not across the land corridor, then across the greatly reduced seawater volume of the Gulf of Mexico. The migrating biota, including humans, as well as natural forces were vectors for dispersal of a broad range of terrestrial vegetation.

Water surely must have been in more limited supply on the Floridan Peninsula, and what few streams and springs there might have been, as well as cenotes (standing water-filled holes in the substrate formed by acidic dissolution of the limestone), served both animal and human populations.

Forested regions in which oaks and other hardwoods predominated were less extensive than at later times, and more savannas and tropical and subtropical grasslands with scattered trees were present. With lower sea levels, obviously the coasts were much farther out than they are today, and during glaciation the peninsula doubled in geographic size. Proof of the larger land mass is substantiated in an example cited in 1994 off the coast of today's South Carolina. The site is about 15 miles offshore, under 55 feet of water, and has eight stumps of what are believed to have been cypress trees. Samples from this offshore site have been radiocarbon–dated back to 10,000 years ago, or by AMS to about 11,350 calendar years ago. Recovered pollen grains and fossils from the sediment are being studied to help determine types of vegetation, climate, and environment of that ancient forest.

At that time human populations were entering Florida and sea level still was as much as 105 feet below present level. Human habitation along the coastal strip back then is doubtful. The strip was, and still is, composed of the ocean, beach, dunes, and the lagoon system. Exposed to frequent violent storms, especially summer and fall hurricanes, and offering few utilizable resources, the sandy, wind-blown coastal strip did not appear to prove favorable for survival of the early people. Since the land mass extended farther seaward when first they ventured into the area, Florida Paleoindian sites also are known to exist below the surface of both the Atlantic Ocean and the Gulf of Mexico, but only a few have been found to date. For example, divers currently report finding Clovis points at sites located underwater in the Gulf a few miles offshore of Tampa.

For that matter, comparatively few Paleoindian sites have been discovered on land, and those that have apparently were left by individuals or small bands. In the more northern part of the peninsula, they often are found at what were kill sites at stream or river crossings. In the southern part of the peninsula, sites more often are found in sinkholes, as indicated by the bones and teeth of humans, their weapons' points, and tools left in association with fossilized bones of creatures that might have lived at the time.

Archaeologists are quick to point out that it is important to understand that mere association of human remains with remains of extinct fauna certainly does not guarantee deposition of both at the same time. A few sites excavated over a century ago erroneously appeared to confirm coexistence of certain animals and humans. One was at Vero Beach, where a human skull and skeleton parts were found with bones of extinct horses and tapir; a second at Melbourne, where a crushed human skull was unearthed in association with bones of extinct horses and tapir. Mammoth and mastodon bones were located nearby. The association might have been assumed to hold true if those faunal bones had displayed evidence of butchering, breakage patterns typical of that to obtain marrow, or evidence of cooking. A stone or bone point lodged in a bone certainly would confirm presence of both a hunter and his prey at some identical place and time. In those two examples, they did not.

Cutler Fossil Site

In 1985, excavation was undertaken at the Cutler Fossil Site, located about eight miles southwest of the city of Miami, within half a mile inland of Biscayne Bay, on what today is known as the Deering Estate. At the time of earliest human habitation, the coastal margin would have been about six miles east of the current shoreline. Situated in a large solution hole,

Right: Excavation revealed artifacts indicating human habitation as far back as 10,000 years ago.

Opposite: The Cutler Site in its natural state typifies area solution holes.

Below: (c) Coral and (d) limestone abraders and (e) the limestone scraper served as tools. (f — two items) The Dalton atlatl points are a type thought to be directly descended from earlier Clovis points. (g) The unusual large limestone point is discolored from an attempt at heat-tempering of the soft stone. (Photographs: Archaeological and Historical Conservancy, Miami.)

the site is located within a hardwood hammock on a part of the Atlantic Ridge, about 15 feet above today's mean sea level, and six feet above the surrounding pine forest. It is the largest of scores of solution holes that penetrate the ridge, measuring approximately 15 by 18 feet at the top.

Collectors already had disturbed the site, so the area was fenced by the property owners, who then invited Miami–Dade County archaeologist Bob Carr and the Florida–based Archaeological and Historical Conservancy to conduct a scientific investigation. An initially planned three month dig ended up a major 14 month excavation project, followed by months of ongoing analysis by a team of interdisciplinary scientists. Included were a physical anthropologist, paleontologists, paleobotanist, geologist, and another archaeologist who also is an expert on Paleoindian artifacts.

Throughout most of the hole, the excavators uncovered a layer of broken and burned limestone rock, within which were found thousands of animal bones, charcoal, and a variety of cultural artifacts including bone fragments of pins, awls and needles, stone choppers, a worn coral rasp, scrapers, and projectile points. Unlike so many South Florida sites, there proved to be a dearth of marine shells. A charcoal sample from this level has been radiocarbon-dated about 7,670 B.C., that is almost 10,000 calendar years old. Of particular interest is the fact that a few of the stone artifacts were made from the oolitic limestone. Archaeologists generally discount oolite as a reliable knapping material because of its softness and unreliable fracture patterns. Attempts to overcome these negative factors were made by some form of heat tempering, as indicated by the naturally white limestone's grey color, which results from its exposure to fire. The Paleoindians were well aware that heating raw materials slowly to critical temperatures for sustained periods, and allowing gradual cooling, made the stone easier to flake and take a sharper edge. However, the peninsula's marine origin formed little in the way of stone of sufficient density and hardness for conventional tools and points, with the exception of chert.

Chert is a class of rock that includes flints, chalcedonies, jaspers, and argillites. It is formed in fossiliferous nodules within ancient marine limestone deposits, and was utilized by the Paleoindians as a suitable substitute for flint and other workable stone. The chert more than likely had been chipped into smaller, lighter to carry preforms prior to being transported from quarries over two hundred miles north of the Cutler site. Somewhere along the way, the preforms had been worked into finished drills, scrapers, and projectile points. They represent a chronological age range of about 6,500 to 5,300 B.C.

The projectile points would have been used on throwing spears, or 'atlatls,' the latter an Aztec word written into the chronicle of the New World by the conquering Spaniards. An atlatl is a short wood or bone shaft with a handle or one or two finger holes on one end, and a hook that fits into the butt end of a shafted spear at the other. It artificially lengthens the thrower's arm, thereby permitting him to propel the spear with greater velocity (up to 500 times the striking force) than that generated by muscle strength and length of the arm alone. The earliest stone or bone spear points were rather large and heavy by comparison to much later, smaller arrowheads and were made without barbs. They appear to have been deliberately designed to fall out of the wound, whereupon the weapon could be recovered for reuse. Others had the point affixed to a short shaft, usually of wood or bone, that was inserted into a socket in the end of the spear shaft. Upon striking an object, the longer shaft, designed to disengage from the shorter one, bounced back in opposite reaction, enabling the hunter to recover it for reuse. The short shaft and its hafted point remained in the wound. Carr reports that results of the excavation in 1986

> indicate possibly one intensive or several short periods of occupation over a short time. Fires were built to cook food, and evidence of tool use and manufacture are prolific. The extensive amount of faunal bones associated with and just below this level, many from extinct fauna, raised the question of whether man was feasting on extinct horse, peccaries, mammoth, bison, or any of the over fifty species identified. We attempted to test the hypothesis that the human habitation occurred concurrently with the extinct fauna. This was done by identifying and analyzing the burned and charred bones, based on the assumption that these burnt bones resulted from food preparation.... Analysis revealed a predominance of extant species, mainly deer and rabbit and even some domestic dog. Extinct species were represented in less than 1 percent of the burnt bone collection. Burnt and charred elements of extinct fauna included bones and teeth from horse, peccary and the giant tortoise. Among these, charred Camelidae elements were uncovered in an undisturbed context 20 centimeters below the rocky cultural horizon — a tantalizing suggestion of man's visit to the Cutler site prior to the intensive habitation of ca. 8000 b.c. Another viable argument to explain the burnt extinct faunal bones is that the late Paleo-Early Archaic period of habitation was situated directly on the bone bed in contact with exposed extinct faunal bones and that some of these were burnt during fires constructed on the floor. This is certainly possible for some of the bones, but the highly charred Camelidae vertebrae and Equus teeth are not what would be expected from the relatively low temperatures created by fires at ground or below ground levels.
>
>   A second explanation is natural fires, but a solution hole beneath the forest floor offers a generally poor burning opportunity because of higher humidity levels. A final alternative explanation is that a relict population

of extinct fauna survived to the time of Cutler's major occupation. However, the accumulating scientific data from many sites in North America indicate that, except for possibly the peccary and horse, most megafauna had disappeared by about 11,000 years ago.

An inspection of some of the extinct animal bones ... revealed no evidence of culturally caused breakage that would be expected from butchering. There can be no doubt that the presence of large numbers of extinct animals at the Cutler site is a result of events ranging from animals naturally trapped or having fallen into the hole, predators using the hole as a den and bringing prey there, birds of prey feeding or nesting, or some smaller species such as toad and small animals probably living within the hole. Man's contribution to the bone debris of the Cutler site represented a minor flicker of time in the long history of natural events occurring there.

Most significant among the Cutler discoveries are three adult and two children's skeletons recovered in an area of deep soft sediment under the southern ledge. Carr continues:

All of the bones were disjointed and broken; some were burned. It is probable that they represent intentional burials, but most were disturbed by various activities as witnessed by the ulna with carnivore teeth marks and breakage patterns, suggesting that some animal had scavenged the body while still fresh.... It appears that all of the human remains date from the period of ca. 8000 B.C. represented by the habitation and cooking activities uncovered throughout the hole. It is probable that these people did not force any dire wolves from their den or find themselves in a dangerous niche somewhere on the food chain above the bison and below the jaguar or dire wolf. More likely, when these people arrived to occupy the Cutler solution hole they found a vacant rock shelter, possibly with spring water and a rock ledge overhang, all of which made ideal protection from the elements. By this time, knowledge of the mammoth and the horse may have already faded into folklore, but it is possible that the final hunt for the last of Florida's peccaries or some lingering megafauna was strategized over Cutler's campfires.

Other animals found in the Cutler site include the jaguar, American lion, bison, sloth, and sufficient quantities of wolf remains to lead one to believe that at some time it might have been a wolf den. Paleontologists on site hypothesized the den was last used about 11,000 B.P.

Monkey Jungle

A few miles southwest of Cutler is a modern tourist attraction called Monkey Jungle. Silt removed from one of the limestone solution holes on this natural 30 acre hardwood hammock yielded bones and teeth of some 50 species of late Pleistocene animals. Many of those were the species found at Cutler. There were peccaries, the armadillo-like Glyptodont, llamas, and horses. Other bones and teeth indicated that some had been hunted by large carnivores such as short-faced cave bears, dire wolves that stood to six feet high at the shoulder, saber-toothed cats, and the huge American Lion, *Felis atrox*, like the one from Cutler. Modern African lions may reach a weight of about 390 pounds. This extinct species of cat is believed to have reached a weight in excess of 500 pounds. While for a time this lion was endemic to broad areas of the North American continent, previously it had been found on the peninsula only as a single skull in a north Florida river. The Cutler and Monkey Jungle specimens confirm that it did roam the entire Floridan Peninsula. The solution hole is thought to have been a natural trap, rather than habitation site of the creatures whose remains were found therein.

A few sand-tempered pottery shards and bone points reveal the presence of late Archaic or early Formative Period people who may have utilized the site around 2,000 years ago. Some small, possibly human bones, and a spatulate human tooth of typically Mongoloid feature (the major ethnic division of the human species characterized by slanting eye folds, low-bridged nose, smooth, straight brow, and prominent cheek bones) retrieved from the muck causes one to reflect on the plight of the individual. Did the human die and his fellows merely dispose of the body in the hole with the rest, or did he fall in while running through the area, possibly being chased by a huge wolf or giant cat? Data gleaned thus far from the Monkey Jungle site cannot resolve any of these hypotheses; it is left to the reader to speculate on the difficulties of survival in Paleoindian times.

Other solution holes within the hammock have yet to be examined. Whether or not the site was one of continuous occupation since Paleoindian times, or more of a temporary glades haven camp from time to time, is unknown. A continuing study of life on the hammock may, in years to come, fill the void between Paleoindian and modern times.

Currently, the attraction invites school groups to investigate the archaeological site for themselves. Silt and material originally removed from the solution hole are placed in small dry plastic children's pools, and students are encouraged to dig and screen to their hearts' content. Artifacts found are returned to monitors, to be included in the Jungle's educational archaeological display. To ensure student success and generate greater interest in the subject, the silt might even be salted with ancient teeth, bone, and the like previously screened out by others. A Monkey

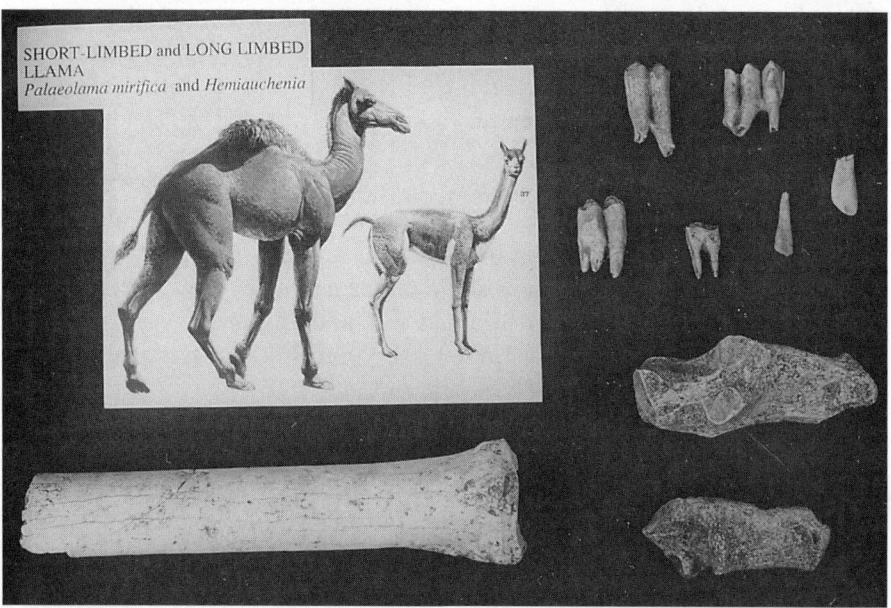

The Paleoindian Period

Above: Volunteers of the Archaeological and Historical Conservancy patiently screen endless yards of solution hole sediment. Bones, teeth, and shell help indentify fauna of the time. Cultural materials such as shell pounders and burned bone help establish human presence.

Opposite—Top left: Kim Davis moves from a small side chamber into the main solution hole. *Right:* At the Monkey Jungle, Miami–Dade County Archaeologist Bob Carr inspects one of six major solution holes. *Bottom:* Displays of artifacts of llamas, short faced cave bear, armadillo, peccary, dire wolf, and horse are to remain at the jungle in a special educational site museum.

Jungle Archaeology Day is held each March, during the state-wide Archaeology Week, at the height of the tourist season. Under tutelage of the Historical and Archaeological Conservancy and the Archaeological Society of Southern Florida, the event is proving extremely popular with educators and students of all ages, as well as tourists visiting the attraction from different areas of the world.

WARM MINERAL SPRING

A third major discovery of earliest human existence in South Florida was made in another solution hole named Warm Mineral Spring. In this case, the site is a water-filled spring located about one mile off Highway 41, 20 miles south of the city of Venice, near the southwest coast. The hole measures about 240 feet in diameter by 230 feet deep. Warm water enters the spring through a subterranean passage at the bottom, averages 87 degrees Fahrenheit in temperature, is heavily mineralized, strong in hydrogen sulfide, and, with the exception of the top 15 or 20 feet, is devoid of oxygen. Below that shallow depth, organic matter cannot readily decay.

Rising from the bottom to the 124 foot level is a huge cone of debris consisting of everything that ever has fallen into the spring over the millennia, beginning with fragments of the limestone that roofed the area, and then collapsed perhaps 20,000 years ago. Bones of Ice Age land mammals, a saber-toothed cat, and giant ground sloth have been recovered, as has a hickory nut. Today the nearest hickories grow 200 miles to the north, indicating a cooler ancient climate at the spring. A giant, five-inch-long shark's tooth and fossil dugong (related to the manatee) bones date from perhaps 30 million years ago when this part of Florida was under the sea. Within the debris cone are coprolites, fossilized waste of animals (or humans). They tell scientists what animal produced them, what they ate, and hold sediments containing pollen that may provide data regarding prehistoric plant life. Analysis of pollen also helps determine environmental changes, if any, via variations in botanicals over time, and evidence of possible agriculture.

In the side walls of the spring, beneath the current water surface at both the 90 foot level and the 43 foot level, are cavities filled with jagged stalactites. Ground water levels are tied to sea level so, obviously, sea level was much lower back then. The cavities had been dry for thousands of years since stalactites only form in air from continuous dripping of mineral-laden moisture percolating through the ground. Geology professor Dr. H. K. Brooks of the University of Florida estimated that sea level had to be

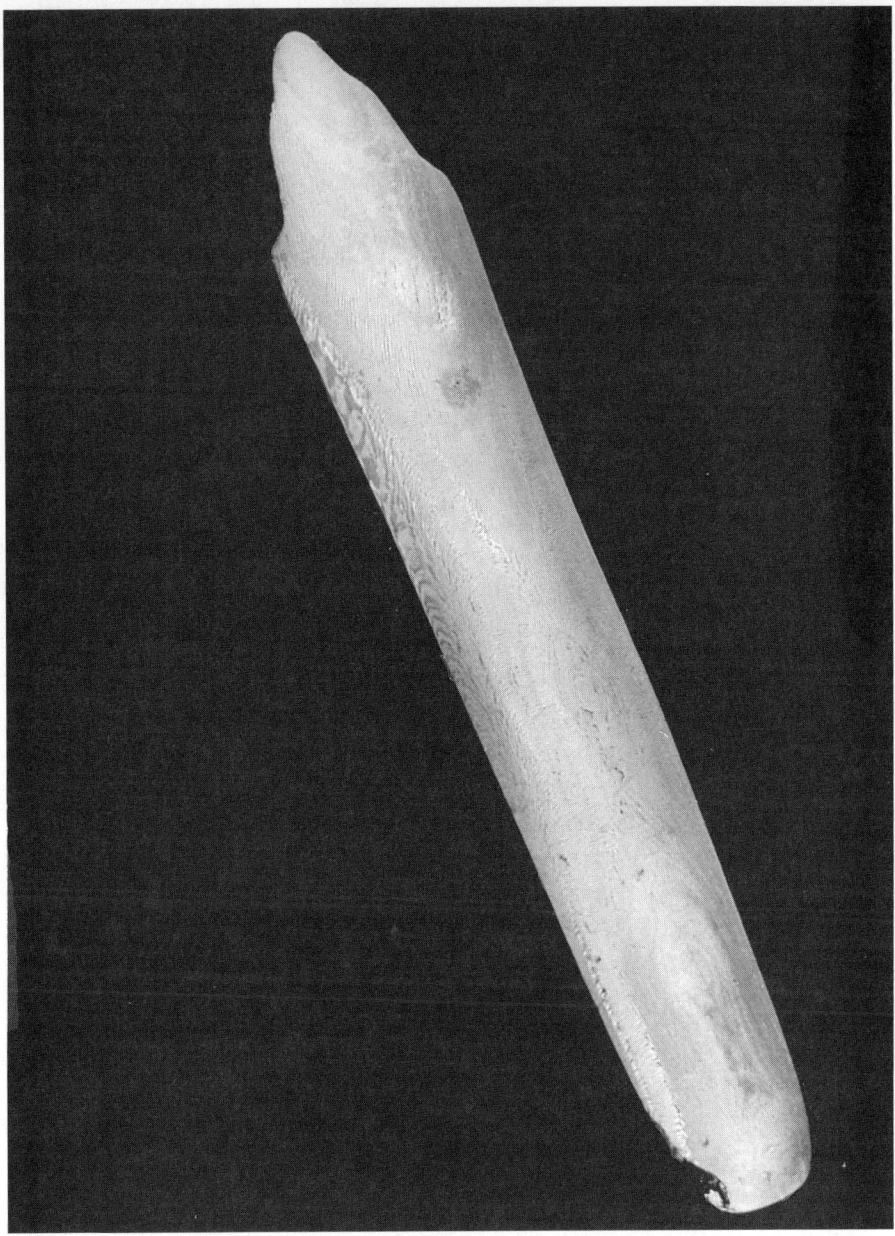

A shell atlatl hook found in Warm Mineral Spring has retained its polish and craftsmanship throughout the millennia. (Photograph courtesy of the Florida Department of State, Division of Historical Resources, Bureau of Archaeological Research, Tallahassee.)

low enough long enough for the caves to be dry and for stalactites to form minimally 6,000 years ago.

The inside walls of the spring are steep, but access to the cavity ledge apparently was possible. When several broken stalactites barring the front of the 43 foot level cavity were removed, they revealed a nearly complete skeleton of an adult male Paleoindian. The two pieces of stalactite had been wedged across the opening of the burial cavity, possibly to deter disturbance by marauding animals. He is estimated to have measured about five feet four inches in height and 110 pounds in weight. He seemed to have been purposely buried in the fetal position often used by early cultures. With him was a carved, highly polished shell end of an atlatl. Incredibly, it still bore faint traces of the dark adhesive that had bonded it to the spear thrower. His skull type more closely resembled those of the Pleistocene beds in Vero Beach and Melbourne, rather than skulls from Archaic Period Indians of 9,000 to 4,000 years ago found throughout the southeastern United States. Comparative analysis of traits of a human mandible taken from the spring, as well as wear and morphology of the dentition, concluded that the man was likely to have been 32 to 36 years old. He is judged to have displayed typical Mongoloid features. There is no doubt that he lived as a hunter and gatherer between 10,500 and 7,200 years ago. These data, combined with height and weight listed above, begin to provide insight into a physical description of at least one of the earliest Floridians.

To date, skeletons of 20 separate individuals have been discovered, including a nearly complete human vertebra and a pelvis fragment carbon 14–dated at the University of Texas to be between 8,260 and 8,630 years B.C. Multiple bone samples tested at several different labs showed a C14 age range between 8,000 to 10,310 years; that is about 11,000 calendar years old (cal B.P.).

Found in layers of clay in the 90 foot cave were several bone needles and other bone artifacts, a shaped piece of antler, wood and charcoal fragments, bird, animal, and the human bones.

Warm Mineral Spring, containing its incredible trove of archaeological material, is listed in the National Register of Historic Places.

Stalactites referred to in the foregoing are found in numerous underwater solution holes. Warm Mineral, Little Salt, and other springs have filled with fresh water. Others, such as those at the Cutler and Monkey Jungle sites, are impacted with humus, earth, and rock. In the Bahama Islands are many filled with the crystalline clear, sapphire water of the sea. They are referred to as blue holes. Some of the internal caverns within any of the holes have been found to contain stalactites and stalagmites. In one on the eastern fringe of Andros Island, divers photographed these magnificent formations at a depth of 180 feet. All attest to shallower seas millennia ago when an incredible volume of the planet's water was frozen in ice.

Additional proof of lower seas is found 20 feet deep off Venice on Florida's southwest coast. Protruding from the Gulf's muck and peat bed are tree stumps and roots, remnants of an ancient swamp. Among them were found ivory, teeth and bones of mammoths and mastodons, bones of the giant ground sloth, fossil beaver teeth, and extinct land tortoise. Among those artifacts of terrestrial animals, to confuse the issue, were found manatee bones, whale vertebrae, and thousands of huge triangular teeth of *Carcharodon megalodon*, the giant shark believed to have become extinct by 1.8 million years ago.

LITTLE SALT SPRING

Another Paleoindian site about three and three-quarters miles south of Warm Mineral Spring in southwest Florida near Charlotte Harbor, Sarasota County, is Little Salt Spring. Today the spring is five miles from the Gulf shore. Ten thousand years ago, lower sea level would have placed the site some 60 miles inland from the shore. The very weak flow from the spring of virtually oxygen-free, heavily mineralized water is apparently a phenomenon linked to present sea level. Evidence of water depth fluctuations within Little Salt Spring so closely parallel those of Warm Mineral Spring, it reinforces the probable change of both from either empty solution holes or still water cenotes to flowing springs about 8,500 years ago.

The aquifer was being recharged by increased precipitation on the north end of the peninsula. All of this water increased pressure within the subterranean artesian system, causing it to seep through substrate fissures into the sinkholes, turning them into flowing springs. The inaccessibility of the spring's location had kept its archeological treasure protected for 12,000 years. Additionally, the sinkhole that forms the spring, which is about 75 feet in diameter and 150 feet deep, is hidden beneath the surface of a much broader lake.

As previously noted, 12,000 years ago the cooler climate and scarcity of fresh water created drier, almost semi-arid conditions rather like that of a modern African savanna. Man as well as animals he hunted were attracted to sinkholes that served as potable water reservoirs. The water level in the spring was then at what today is the 90 foot level underwater, and it was on a ledge there that the collapsed shell of an extinct species of giant tortoise was found. Embedded in the four-foot-long shell were fragments of

Top: The area of Florida's Little Salt Spring is pocked with solution holes, each a potential preserve of the past. (Carl Clausen photograph.) *Bottom*: The lab adjacent to the spring is utilized by the University of Miami's Underwater Archaeological Research Team under direction of Professor John Gifford. (Photograph courtesy of the Florida Department of State, Division of Historical Resources, Bureau of Archaeological Research, Tallahassee.)

a wooden red mulberry stake that possibly had been used to kill the tortoise. Beneath it, bits of fire-hardened clay indicated that a fire had been lighted to cook the reptile in its own shell. Radiocarbon dating of the stake places it at 12,030 years ago. Other bones found along the ledge included those from the turtle, diamond back rattlesnake, rabbit, wood ibis, an extinct ground sloth, mastodon or mammoth, and bison. Human remains at that 90 foot level eluded the archeologist's search, but if ever they are found they very likely would prove to be among the oldest discovered in North America.

By about 8,000 B.C. the water level had risen to approximately 31 feet below the present surface. The Paleoindians lived on the slope of the basin around the sinkhole. There they built their fires, prepared and cooked their game, principally white-tailed deer, and worked with their wood, bone, stone, and shell materials as attested to by those artifacts being found in association with drowned informal hearths on the sand surface below more recent sediments now lining the basin.

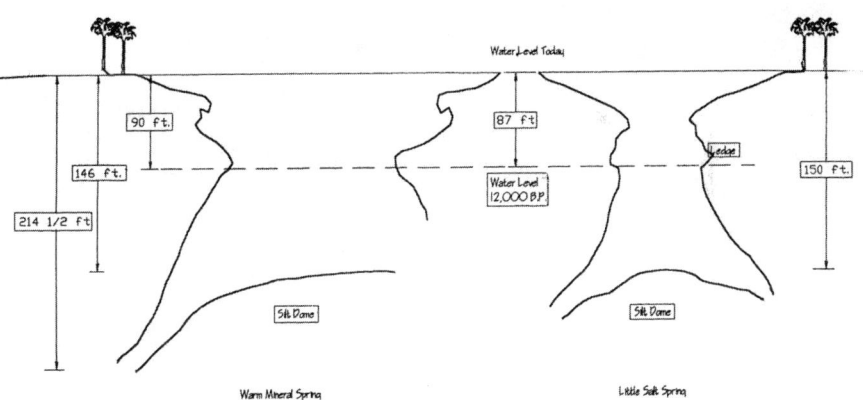

The sketches above compare Warm Mineral Spring and Little Salt Spring in cross section. The ledge indicated at the 87-foot level of the latter is where divers found remains of a large tortoise that had been killed with a sharpened stick and cooked in its shell. Radiocarbon dating of the fire's charcoal indicated a date of approximately 10,000 B.C. A ledge at the 43-foot level of Warm Mineral Spring held stalactites, behind which were Paleoindian remains dating back to 8,000 B.C. Artifacts have been recovered from the upper slopes and from the silt domes of both. (Not to scale.)

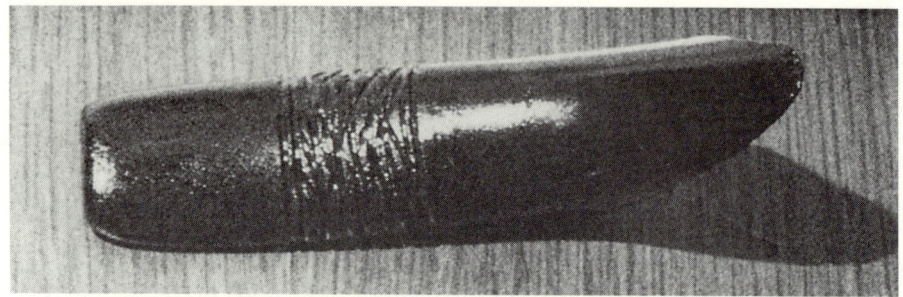

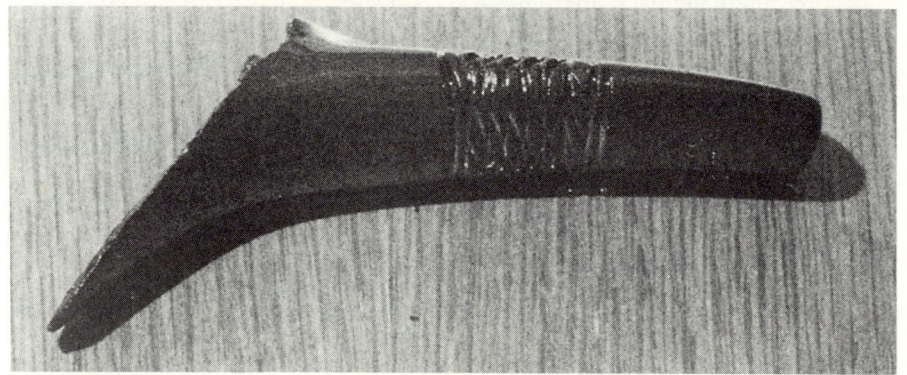

Little Salt Spring artifacts include deer antler atlatl handles incised with geometric patterns.

A series of crudely pointed stakes dating back to 7,720 B.C. had been thrust into the ground near the edge of the water. It has been speculated that they formed a barrier to assist the Indians in driving game into the water, wherein they could easily be killed. The sediment into which the stakes had been driven also yielded hickory nuts dated at 7,920 B.C. Pollen samples from the same sediment layer showed that the dominant trees at that time were wax myrtle, oak, pine, and hickory. The only herbaceous plants were ferns, grasses, and composites (a vast botanical order including plants like the daisy, thistle, dandelion, and others that constitute almost half of the flowering plants of tropical America). This suggests that the slope was well-drained and the surrounding area relatively dry.

A socketed antler projectile point with the wooden tip of the shaft preserved in its base, and the bottom portion of a carved oak mortar dated at 7,080 B.C. also were collected. A non-returning oak boomerang retrieved from the depths was similarly radiocarbon-dated. Before the recovery of this specimen, evidence for the use of this particular type of weapon was

The Paleoindian Period

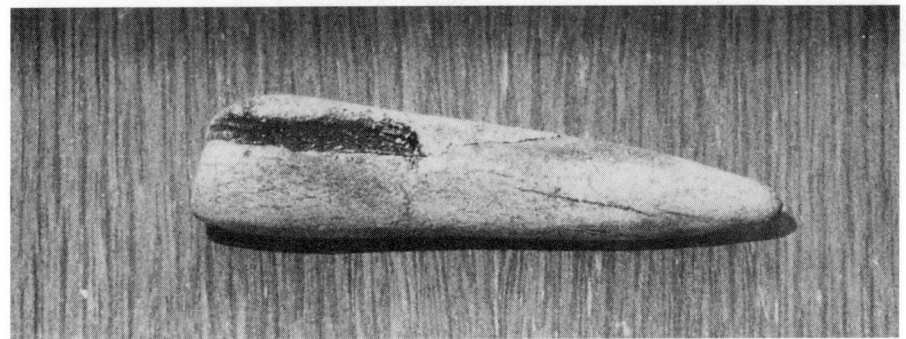

A socketed antler projectile point.

A composite hammer of antler, the oak handle of which still was undergoing conservation at the time the photo was taken.

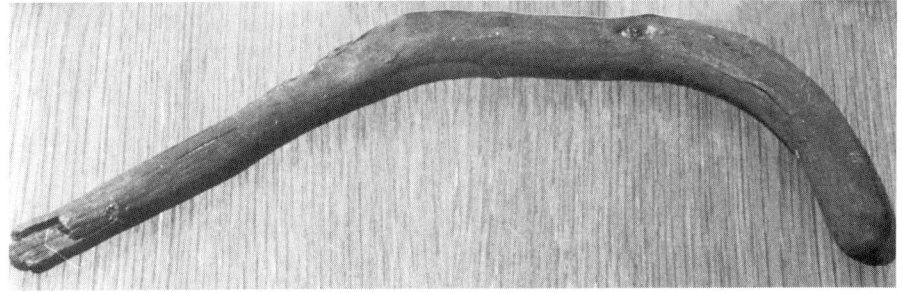

A strongly curved ten and one-half inch oak stick with a blunt tip, the other end of which is broken, function unknown.

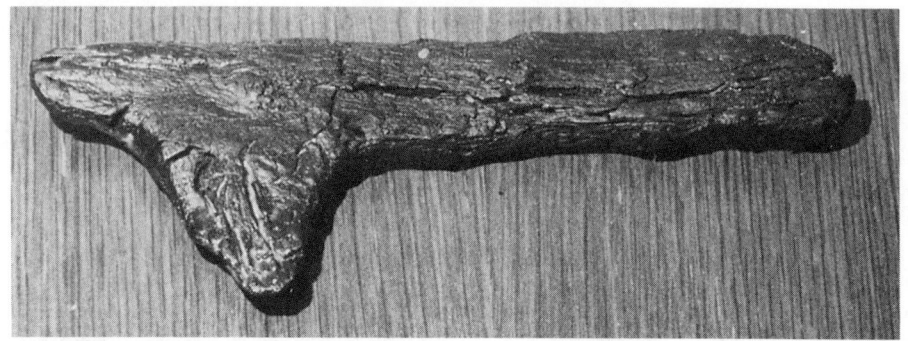

A six and one-half inch section of a boomerang or throwing stick.

A six and three-quarter inch handle of a wooden dipper or spoon.

The base of an oak mortar carved from the section of a small tree trunk six and three-qurter inches in diameter.

Fragments of a carved wooden throwing stick.

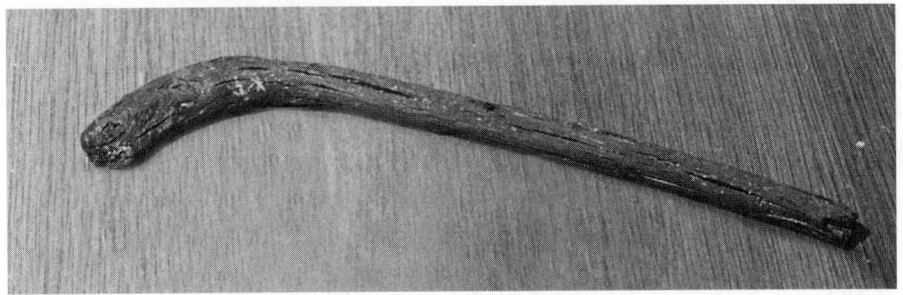

A small clubbing or throwing stick of white oak.

A sharpened stake of red mulberry with a fire–hardened tip. Found at the 87-foot level between the plastron and shell, the stake is thought to have been used to kill the giant tortoise. It was radiocarbon–dated at 12,030 years ago (10,030 B.C.). All other wooden artifacts were dated back to about 9,000 years ago. (Courtesy of John Gifford and Marty Healy, Rosenstiel School of Marine and Atmospheric Sciences, University of Miami. Photographs by Ed Thompson.)

found in Australia, in ancient Egypt, where specimens were among the grave artifacts of Tutankhamen, in India, and in western Europe. Modern laminated wood models of this type of boomerang displayed excellent stability in flight, and were judged fully capable of maiming and downing game up to the size of small deer at a range of 50 to 60 yards. This oldest known specimen of true hunting boomerangs refutes the earlier belief that they were unknown in the New World.

Also recovered was a portion of what had been a gourd vessel with a bail hole near the lip. This was dated from about 7,080 B.C. to possibly 6,000

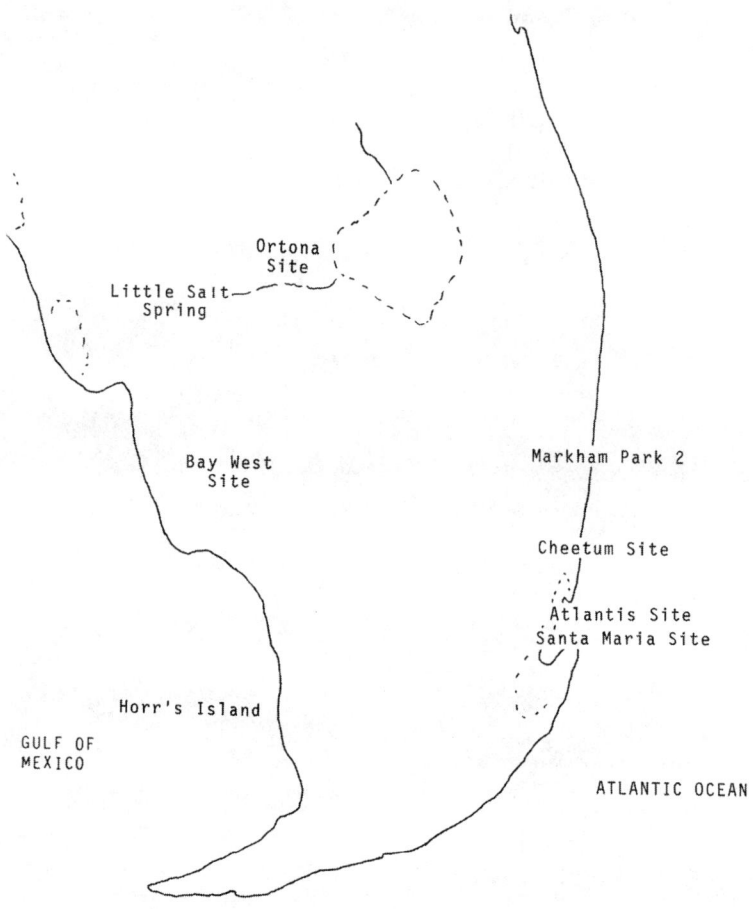

The Floridian Peninsula of 7,000–500 B.C. displays the alteration of the shoreline caused by rising seas. The major bays are forming and the great Lake Okeechobee fills and discharges through rivers east and west. To the south the earth becomes wetlands as far back as 5,000 B.C., precursor to today's Everglades. (Not to scale.)

B.C., making it the oldest artifact of that type found in North America. Remnants of the gourd-like form of squash, *Cucurbita pepo*, and the bottle gourd, *Lagenaria siceraria*, are found in sites virtually throughout the peninsula in context from Paleoindian to modern times.

The water level in the spring and throughout the Floridan Peninsula rose steadily for several millennia as the climate warmed. By 6,500 B.C., water from the cenote, now turned to flowing spring, invaded the drainage leading into the basin from the northeast and created conditions favorable for deposition of a brown, sandy peat. This peat layer contains a pollen assemblage similar to that of the modern Everglades sawgrass-water lily habitat of grasses and composites mixed with water lilies, cattails, and arrowhead. Dominant trees were small amounts of oak, pine, and willow.

Over the span of thousands of years animal and human predation plus the continually changing climate had eliminated remnants of the huge animal species of the Paleoindian Period. Paleoindian utilization of the spring apparently came to an end with this high water table and the increased availability of fresh surface water throughout the area. Paleoindian numbers were few, especially throughout the distal end of the peninsula, but they were here, and now were more free to roam.

A major frustration for modern archaeologists must be the known location of additional solution holes south of Lake Okeechobee that, more than likely, hold artifacts and remains that might fill gaps in the ancient puzzle. Some of these are Deep Lake, Tarpon Lake, Rock Lake, Lost Lake, and Lake Sampson. Each is about 90 to 110 feet deep and is fed fresh water through underground streams. Originally they were much deeper, but floods from the great lake to the north have partially filled them with large quantities of silt, through which researchers would have to excavate deep underwater to reach the most ancient artifacts. Research is further complicated by the wilderness locale, suffocating heat, hordes of mosquitoes, vicious biting flies, spiders and scorpions, poisonous coral snakes, rattlesnakes, and water moccasins, and the fact that these lakes are havens for vast numbers of alligators, all of which is sufficient to dampen the enthusiasm of the most dedicated scientist.

3

The Archaic Period: 7,000–2,000 B.C.

Changing climate permitted a wide variety of plants and smaller animals, deer in particular, to flourish on the Floridan Peninsula. Aboriginal nomadic movements reduced as water and food sources became more available. The people became more efficient in exploiting a broader range of environmental resources. The Paleoindians were entering a transitional period from the lingering nomadic, megafauna hunting and gathering way of Ice Age life to subsistence hunting, trapping, fishing, and gathering as demanded by ecological adaptation.

In due course, their numbers increased as life became more sedentary and migrations became more seasonal in nature. They might have been forced to choose to relocate in response to fluctuations in sea level, availability of freshwater, reduction or loss of food supply, conflict, pestilence, or any other number of variables.

Aboriginal tool making became more sophisticated. Although the people continued to use the spear thrower (atlatl) as one of their primary weapons, and maintained other cultural affinities with their forebears, they differed enough in their way of life from the Paleoindians for archaeologists to term this major variation in behavior as the beginning of Florida's Archaic Period of human development. The time frame of this stage would last from about 7,000 to 2,000 B.C.

At some point in time after the beginning of the Archaic Period, reduced availability of surface freshwater again restricted nomadic movements. Pollen samples of that time suggest that xeric conditions in South Florida might have been severe enough to damage botanicals and force the small population of humans, and the animals they hunted, into a northward migration. The huge freshwater lake, to be known first as Mayaimi and later Okeechobee, had not yet begun to fill, nor were there brackish estuaries due to lack of freshwater runoff to the sea.

Weston Pond

Weston Pond is a large circular marsh and solution hole fairly recently discovered by archaeologists. It is located inland of Ft. Lauderdale and the peninsula's southeast coast.

Radiocarbon dates from the bottom of Weston Pond suggest that the peninsula's unique wetlands actually may have begun to form by 5000 B.C. That is 2,000 years earlier than previously thought based upon radiocarbon dates that had been secured across the Everglades during the past 30 years. The new evidence reveals that another mesic period of wetness had evolved by 3000 B.C.[1] or earlier. It was during this 2,000 year period that sea level rose approximately 16 feet.

The huge lake, Okeechobee, did begin to fill and, in time, rivers and streams began to flow, forming estuarine and coastal marsh environments where they met the Atlantic Ocean to the east, the Gulf of Mexico to the west. Just offshore sea grasses flourished, and this new environment proved to be healthy for nurseries for a multitude of species of fish and invertebrates.

In addition to the great Lake Okeechobee, by about 4000 B.C. six lakes formed in a Gulf coastal prairie area to the northwest. Rising sea level coupled with a settling of the land had moved the Gulf shoreline inland to about eight miles from that of the present day and Archaic campsites that once were located on the lakes' shores were inundated. By 2000 B.C. the six lakes and rivers had coalesced into one and, in due course, joined with the sea to form Tampa Bay. The lake depressions and those of the rivers that had connected them are evident under the bay today.

Little Salt Spring

The aboriginals of Little Salt Spring dispersed, abandoning the area as a major camp. Rising water filled the spring's basin and formed a slough to the northeast. Layers of sediment deepened to form a rich peat at the bottom of the slough that would act as a natural preservative. Thereafter, the people returned to settle, at least seasonally, on the higher ground parallel to Little Salt Spring's slough.

Archaeologists believe they lived in family groups of 20 to 50 persons.

---

[1]*5000–4000 B.C.: earliest cities in Mesopotamia are forming. Egyptians develop their calender regulated by the sun and the moon consisting of 360 days divided into 12 months of 30 days each.*

52   A Prehistory of South Florida

Over the centuries more than a thousand of their dead were buried, along with countless artifacts, in the protective peat. As the water receded and the soil at the upper level of the slough dried out, the burials were made at steadily lower levels where the peat was still soft. The bodies were extended and sometimes wrapped in grass. Artifacts interred with the dead were almost all utilitarian. Digging sticks, deer bone points, and tools of bone and shell were placed with the people, perhaps for use in the afterlife, perhaps simply interred as treasured possessions of the departed.

Associated stone projectile points discovered along with the foregoing proved to be the large, triangular, stemmed, highly distinctive type known as Newnan's Lake points after their type site in north central Florida. Specimens of this easily recognizable point have been found in archaeological sites throughout the Floridan Peninsula. Their presence suggests that the Archaic inhabitants of the spring were part of a growing population that strongly influenced much of Florida.

Most of the artifacts were well preserved, including one skull containing traces of convoluted brain material that dated back to about 4,000 B.C. However, DNA tests of that brain tissue presented an enigma; it contained

Newnan Lake atlatl points or knives are named for their type site in North Central Florida. They date back to the Archaic Period and are found in archaeological sites throughout the peninsula.

a previously unknown mitochondrial DNA sequence. This suggests that in addition to the three known lineages that had migrated into North America, perhaps a new group of humans had come in as well, but had died out at some unknown point in time. The youngest radiocarbon date for a human bone specimen, which was recovered from about 27 feet below the present water surface of the basin, was 3,220 B.C.[2]

Cypress swamps and hardwood forests characteristic of the subtropics began to develop and the Archaic people were able to abandon the spring for the last time. Seasonally, now the gatherers would find acorns and other hardwood nuts, roots, fruits, berries, seeds, and other food more available. The more numerous human remains and greater quantities and varieties of associated artifacts discovered clearly indicate formation of increasingly larger social groupings, at least at certain seasons of the year.

This was a period of environmental stabilization. There was a slowing of sea level rise; a broad sheet of freshwater flowing slowly from the great lake southward began to saturate the southern limestone topography. Eventually, it became known as Shark River Slough. It was, and still is, the primary discharge for Lake Okeechobee, flowing southward in a 40-mile-wide sheet for 90 miles, at depths ranging from three feet during the summer rainy season to virtually dry, except for its deeper channels, during dry times. The other major, although much smaller slough, Taylor, flows to the southeast and drains into Florida Bay. In the millennia before man's canals altered natural water flow, the wetlands lying south of the great lake covered more than 6 million acres.

In the upper glades and along the sloughs, dense forests of sawgrass (*Cladium jamaicensis*) would grow as high as 14 feet. As the older growth died out and decayed, a bed of rich thick muck formed that often reached a depth of 12 feet. In the lower glades, the hard limestone surface eroded into pockets wherein vegetation could gain only a precarious hold and grasses, even cypress trees, were stunted.

During dry times the humus and peat served as fuel for fires that burned both on and under the surface of the ground. The latter, in particular, posed a serious hazard to both man and beast, as there might be no telltale warning of the smoldering material underfoot. According to John Kunkle Small, (1929) botanist and taxonomist of the New York Botanical Gardens:

> the great sponge of humus, in some places six feet deep, has been destroyed down to the sand foundation by internal fires burning con-

---

[2] *The first date in Mayan chronology is listed as 3372 B.C. The earliest known Sumerian cuneiform writing in the form of wedges on clay tablets dates back to this time.*

tinuously for years.... Half-wild domestic animals live in a half-starved condition, but their numbers have been greatly reduced by fires in the humus. Driven to seek food in the humus regions, which are often afire for long periods beneath the surface, the hogs and cattle fall through the superficial crust only to have their feet burned off in the fire beneath. They consequently perish.

In 1836 a Dr. Strobel reported:

> A gentleman of Cape Florida [the tip of today's Key Biscayne] informed me that he had cut down trees, with the intention of clearing a field to plant. In order to get rid of the trees which had been felled, in an expeditious manner, he set fire to them, and on the following morning was greatly surprised to find that he had not only destroyed the trees, but had also burnt off the whole of the soil, and left nothing but the bare rocks.

Of course, the early humans and the creatures with which they cohabited had to avoid such hazards, especially during periods of drought.

The now continuously flowing, warmer and clearer streams did improve habitats for shellfish. Coastal waters warmed too, and these factors all contributed to development of marsh, barrier island, and coastal lagoon systems. Decaying marsh grasses, microorganisms, and various algae supplied food for a host of more complex life forms. The marsh areas, shunned by modern man, provided an abundance of life forms and sustenance for the low population density of early aborigines. The lagoons frequently are brackish from the constant mix of freshwater runoff and the sea. This provides the necessary sustainable habitat for many species of shellfish, the most prolific of which are oysters. Often open to the sea, the lagoons also are a zone of interchange between freshwater and marine species. Terrestrial plants and animals common to more inland areas usually are available adjacent to, or not far from, both of these productive habitats. This is especially true along the southern Gulf coast's myriad bays and estuaries, making it a particularly rich hunting and gathering area for the Archaic people.

Bay West Site

While dredging peat for landscape purposes in a nursery near Naples, Collier County, about 68 miles south of Little Salt Spring and six miles inland from the lower Gulf of Mexico coastline, a cypress pond mortuary was uncovered dating back to 4,900 to 4,400 B.C.

Artifacts found at what became known as the Bay West Site included five projectile points and a knife blade or scraper. Four of the points had been knapped of coralline chert, several of which had been heat-altered. One point was of fine grained quartzite. The knife blade or scraper had been made of agatized coralline material (quite common in the Tampa Bay area). A large, fine grained, slightly translucent green stone bead of non-local origin was recovered, as was an antler atlatl hook. While at the time of habitation the site more than likely was six miles from the Gulf coastline, it is significant to note that no shell artifacts were found at Bay West. Apparently, Archaic Indians in that location at that time were more terrestrial than marine in their resource utilization.

Millennia later, higher sea level flooded many Archaic sites, including places where chert nodules and agatized coral had been mined for

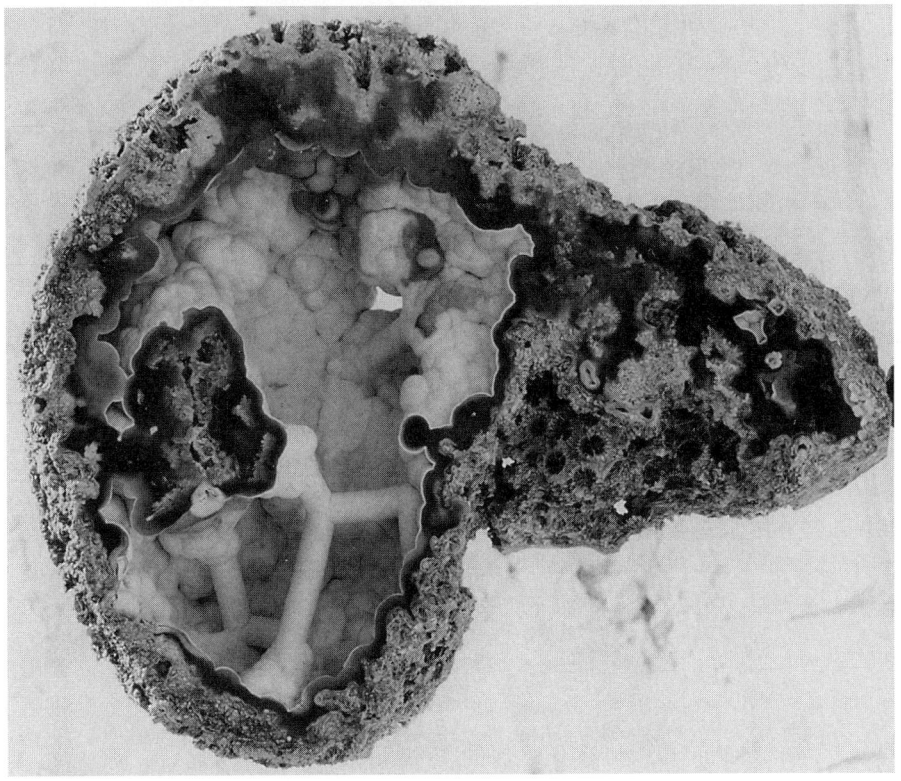

Agatized coral. Tiny exterior polyp skeletons on the stony head reveal its marine coralline origin. Within, minerals have partially agatized the mass. With the more fragile exterior chipped away, the agate served as a substitute for flint, was heat–treated, and knapped into sharp points, gravers, and knives.

knapping of durable stone materials. Generally speaking, there is a notable difference between knapped tools of this period and those of the earlier Paleoindians. There are a larger number and greater variety of Archaic Period tools, but they display lesser degrees of workmanship, some having barely received enough shaping to render them serviceable. Multi-purpose tools seem to be more common, as do large choppers and hammer stones.

Wooden artifacts preserved in the anaerobic peat of the Bay West site include posts or stakes that had been burned at one end, and the charred material scraped away to form end points. Small carved wood shafts retrieved from the peat are thought to be fire sticks, designed to be rotated rapidly enough to create sufficient friction heat to cause tinder ignition.

Partial and disarticulated bone remains of 35 to 40 individuals were recovered from the mortuary pond. They represented both sexes of individuals of varying ages, and some of the skulls may exhibit artificial flattening. Or, this may simply be an effect of prolonged burial. Should the former prove true, it may portend a practice more common later among Native Americans, being accomplished by affixing a headboard to an infant's carrier in a manner to cause the soft skull to grow in conformation with the angle of the rigid wooden plane. Heavy tooth wear was universal and bone degeneration around the teeth due to plaque, abscesses, and subsequent tooth loss was easily noted. Arthritic-like bony growths were clearly visible on the vertebrae.

It is quite clear at this point that a strong association existed between early humans, their burials, and water. Water is thought to have been considered a natural barrier between the living and ghosts or supernatural beings. There are numerous examples of Indian villages being separated from burial areas by water. Therefore, it is not difficult to interpret earliest water cemeteries of south and central Florida as reflecting similar beliefs. Fear of returning spirits was widespread among traditional societies, and discovery of growing numbers of mortuary ponds and solution holes demonstrates that such sites are not isolated occurrences, but may represent at least one type site for the Archaic Period in Florida.

Horr's Island

Three hundred acre Horr's Island is located on the Florida west coast south of Marco Island near Naples, Collier County. Investigations of the island had been undertaken over the past 100 years by a handful of notable archaeologists, but it was not until 1989 that Ronto Developments Marco, current property owners, made possible a full scale study of the site. Archaeologist Mike Russo, an expert on the Florida Archaic, ran the project "which led to a new view of the Archaic people and their lifestyle," according to the report by Claudine Payne in the Florida Museum of Natural History's *Calusa News*. To this point, the accepted model for the Late Archaic Period was that small bands of hunters, fishermen, and gatherers lived in temporary structures and moved seasonally.

Russo's excavations on the island revealed more than 600 postholes, probably representing many small circular houses. Postholes occur when housing support posts are driven into the ground, later to rot away and be replaced by midden material of color different than the ground that falls into the hole from above. Posthole molds permit the archaeologist to estimate the size and nature of platforms or dwellings the posts may have supported. In spite of the conventional view of Archaic people as fishers, hunters, and gatherers, dating of the sites indicated that by 2800 B.C. people were living on the island year after year.

Various animals display clear seasonal growth patterns which are reflected in their bones or shells. Study of such remains found in a site reveals at what season the animals were collected by the ancients, thereby indicating when those people lived there. The thousands of remains scrutinized indicated that "people lived on Horr's Island year-round, gathering scallops in summer, quahogs in winter and spring, and catching catfish, pinfish, and threadfin herring mainly in the fall." The researchers concluded that "the site was a permanent, sedentary Archaic community." The long standing paradigm of Archaic life must be changed, at least in the Caloosahatchee Cultural area of Southwest Florida.

Until this study, archaeologists thought that Archaic people built no mounds at all. That proved not to be the case. The team determined that one Horr's Island conical structure was created of "layers of sand that had been spread carefully over shells from time to time. Some of the sand layers were pure white, some had been colored by the addition of charcoal. The people had built a well-defined cone rising to a point almost 20 feet above ground surface." It was given the designation Mound A. On the outskirts of the mound, Russo found two human burials dating to a slightly later time. Since Mound A is not an irregular midden of shell, sand, and an accumulation of debris, the archaeologists believe it clearly was constructed to serve ceremonial use, possibly as a burial mound. According to Russo, if the latter proves correct, "the mound is the earliest burial mound known in the United States."

In addition to the foregoing, the island does contain later massive shell middens, as well as a contact period Calusa village (see pages 98–99) as do many of the islands and sites within the Caloosahatchee and Ten Thousand Islands Cultural Areas. Horr's Island is discussed here as but one example of a habitation site encompassing millennia of human culture.

Evolution parallel to that of the west coast on the peninsula's eastern shore was minimized, in part by the natural barrier of the Atlantic Ridge. In some places, southeastern streams and rivers managed to cross the ridge and permit gentle rapids or riffs, such as those that formerly existed on the two forks of the Miami River. In other places, the water coursed underground to eventually surface in the form of springs on the eastern side of the ridge, such as those near the Atlantis and Santa Maria sites just south of Miami, Silver Bluff in Coconut Grove, and the Snapper Creek site.

Atlantis Site

On the peninsula's east coast, approximately one and one-half miles south of the mouth of the Miami River, on the Atlantic Coastal Ridge adjacent to Biscayne Bay, high rise construction workers uncovered an ancient

The Atlantis site was found on the Atlantic Ridge south of the mouth of the Miami River. The abundance of broken shells revealed its use by early Tequestas. Note the thickness and large size of the shells that in ancient times were far more robust than today. (Photograph: Archaeological and Historical Conservancy.)

habitation site. Bulldozers and backhoes also uncovered an important prehistoric burial site some 35 yards west of the habitation. Partial remains of six individuals were found below piles of limestone rocks that appeared to have been placed intentionally over the graves. There was no pond or soft peat nearby in which to inter the bodies, as one finds at more inland and west coast sites. Therefore, the rocks may have provided a substitute for supernatural protection, thereby fulfilling the need for the living to maintain distance from the dead via physical obstacle. Or, this may have been done to protect the bodies from predators.

Located on a fairly high and dry limestone bluff paralleling the shore of Biscayne bay, material from this Archaic site has been radiocarbon-dated back to 2,890 B.C. Since Biscayne Bay was a narrow channel prior to 3,500 B.C. and was nearly flooded by 1,200 B.C., both the habitation area and the burial ground apparently were not right on the shoreline at the time of habitation. Today the location is known as the Atlantis Site.

Santa Maria Site

A similar site is located less than a quarter-mile south of Atlantis, the Santa Maria Site. Here, too, the late Archaic people found minimal shelter against the stony hump of the Atlantic Ridge. Freshwater underground streams found their way to the surface of the eastern side of the ridge nearby and flowed as springs.

A campfire-like ring of stones, within which were abundant bits of charcoal and burned animal bones and teeth, leave little doubt of the site's human use. Human skull fragments and a segment of jawbone containing teeth indicate a burial site against the limestone ridge. Archaeologist Richard G. Haiduven commented, "some of the teeth were not at all worn and, so, probably were those of a child."

Among the artifacts uncovered were an oolite grinding stone and pestle of the same material. Similar grindstones have been found at Granada, Snapper Creek, and other sites at the southern end of the peninsula. When the handstone is worked against the grindstone, tiny ooids are dislodged from both, forming a course sandy residue thereon. If the person were crushing and grinding acorns, seeds, or other foodstuffs, that residue would become an inseparable part of the flour. Small wonder, then, that such grit, along with other sandy residue and vegetable matter in the diet, and use of the teeth as a fifth appendage, resulted in the extreme dental wear found in older ancient people.

When aboriginal people abandoned the site is undetermined at this time.

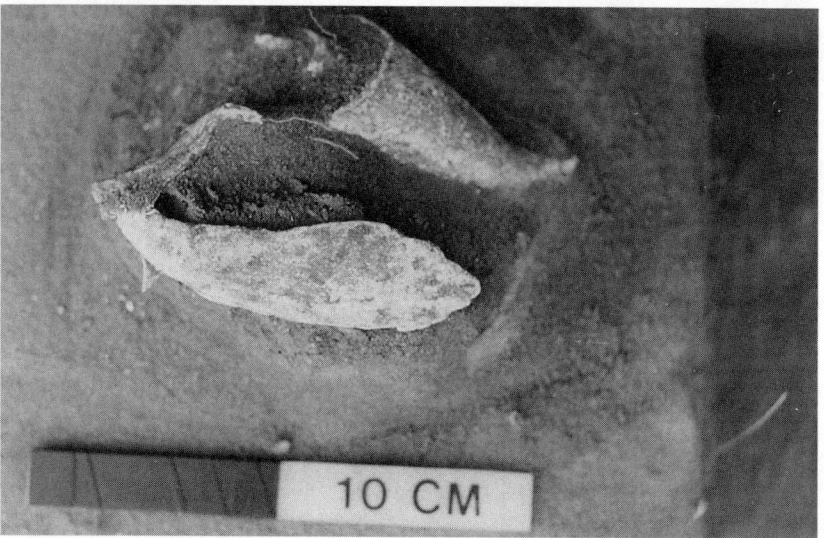

Top left: The Santa Maria site is worked patiently by Archaeologist Richard Haiduven, in spite of the press of heavy construction equipment that soon will cover all evidence of ancient man beneath a modern condominium. Seemingly broken shells are in reality important artifacts, clues to survival in a difficult land. *Right:* The whelk shell outer whorl under Haiduven's elbow served as a dipper or dish. *Bottom:* The Queen Conch is a food remnant and possibly served as some sort of tool.

However, artifacts hundreds of years old of south Florida's first recorded modern European settlers were recovered there. The property has remained in use, with the exception of a time during the Seminole Indian War in the mid–1800s.

Similarities of the sites of human habitation in parallel time frames along the Atlantic Ridge are striking. As was the case with Atlantis and other sites, Santa Maria was dug by archaeologists as backhoes and bulldozers expanded and deepened the footings for a new 51 story condominium tower that has destroyed most of the site, although a small area including a portion of the original bluff has been preserved.

CHEETUM SITE

Cheetum is another site in the east coast's Miami–Dade County, located at its prehistoric time of habitation on what would have been the eastern perimeter of the incompletely formed Everglades. Almost ten miles west of the old headwaters of the Miami River, aerial photographs plainly show remains of an old slough 500 feet wide that bisects the site's hammock area. The slough can be traced southeast to the Miami River, and possibly offered canoe access to the river and, thence to the bay.

The lowest level of the site was radiocarbon-dated back to about 3,120 B.C., while charcoal near the uppermost level was dated about 1,000 years later.

Food remains found in the hammock included crabs, snails, whelks, Queen Conchs, sun ray Venus clams, fish vertebrae, sharks, turtles, snakes, and alligators as well as grey fox, opossum, rabbits, otters, and deer. The only bird remains were the bill of a single white egret.

There were some 21 secondary human burials ranging in age between one and one-half to 50 plus years, with the majority between 15 and 25. The site has been destroyed by rock and sand quarry activities.

MARKHAM PARK SITE NO. 2

Markham Park Site No. 2 is located north of Miami in Broward County, about ten miles west of the intersection of U.S. 441 and State Road 84. The antiquity of the site is thought to be in the late Archaic Period based upon the bottom strata find of a specific type of pottery shard dated from that period from another area of Florida. Artifacts other than pottery were common, including bone, shell, and stone objects in unusually

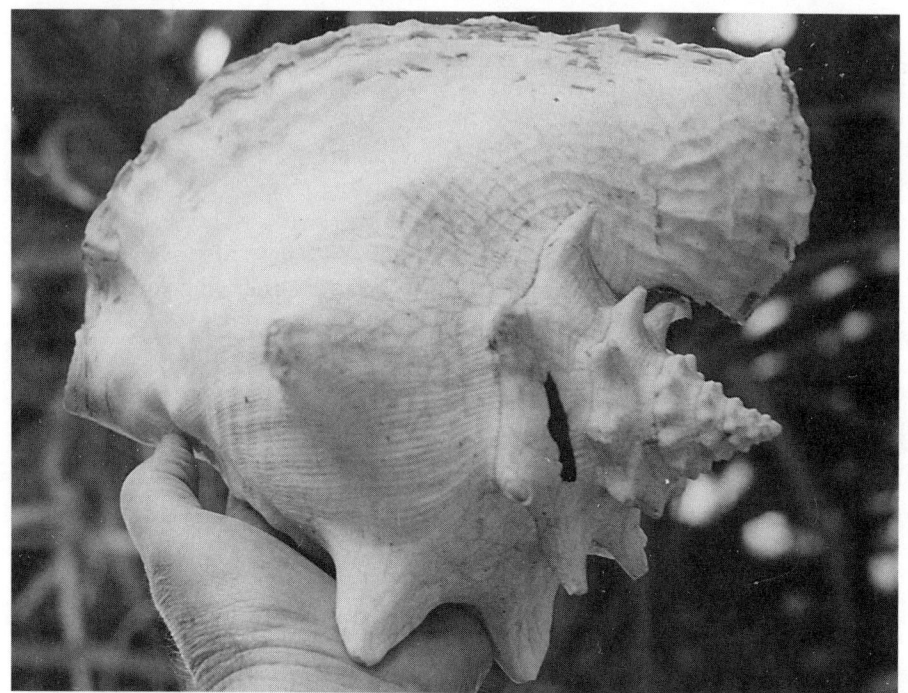

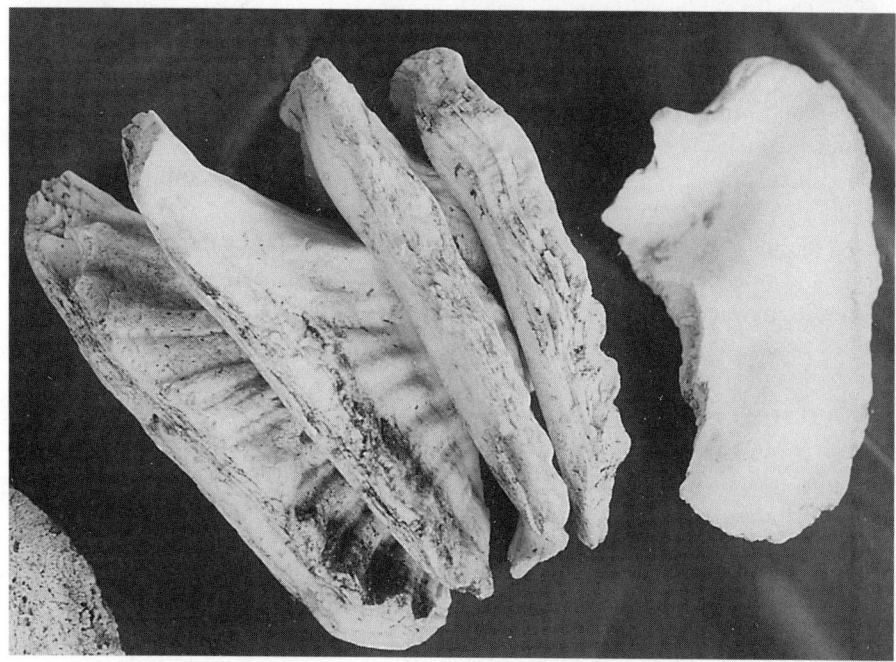

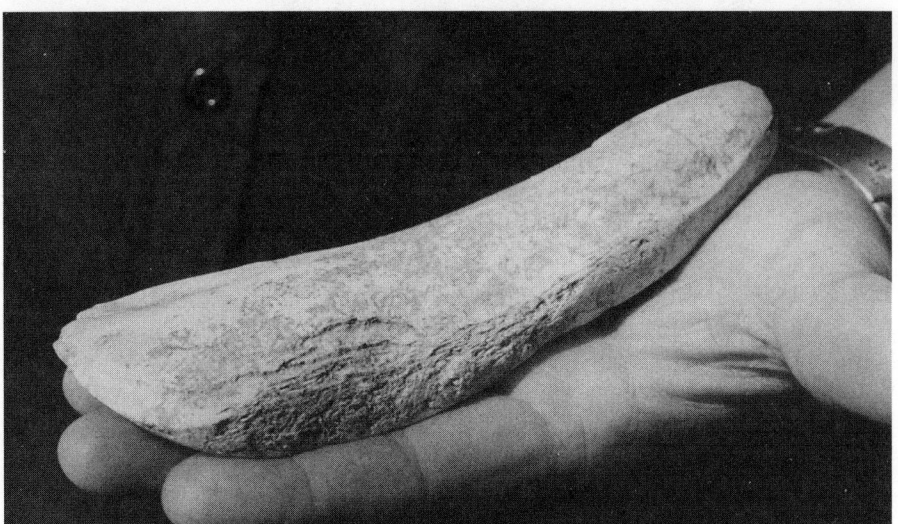

This page: Conch shell celts. *Top:* A celt served as an efficient hand ax. Hafted on a horizontal plane, it was used as an adze. *Bottom:* The columella tip had been honed with a piece of pumice at a sharp angle to serve as a gouge. While the tools might have been inefficient on fresh wood, once the wood had been burned the remaining charcoal would have been removed with ease.

Opposite: The lips from Queen Conchs (*top*) were found (*bottom*) broken away and stacked for future use, probably to be shaped and honed into celts.

high numbers, as were shark's teeth drilled for hafting for use as cutting tools, and shark vertebrae used for ear buttons or drilled for beads.

Most of the unusually large number of 165 shell celts found were in the lowest levels and are believed to have been used in canoe building. The oldest known dugout canoe in Florida (not at this site) has been radiocarbon-dated back to about 3750 B.C. Fire was used to char the wood, and pine trees most often were utilized rather than cypress, as the pine's resin canals aided in the fire-hollowing process. Thereafter, the shell celt made relatively easy work of chiseling out the carbon residue. Numerous canoes have circular burned areas inside on the bottom, indicating that fire was transported with the people as they moved. This particular site seems to have been a stop along a waterway that connected the glades to the ocean.

To the utilitarian list of *Busycon* (whelk) shell tools and utensils, and *Macrocallista* (Venus clam) shell knives, were added stingray spines and a number of pieces of sailfish bills. This reinforces the thought that the early users of the mound were frequent visitors to the miles distant coast.

Most of the aforementioned artifacts are typical of southeastern Florida hammock sites. Points of flint from a higher stratum of the site date back to about 500 B.C., while others from successively higher strata are dated to more recent times. Several points appear to have been knapped in the Paleo or early Archaic Period, but this does not mean that earliest habitation of this site dates from those times. Crude, often reworked points similar to earlier types are found at Florida sites. People simply picked up older points and reworked them as necessary to serve again. Among other stone objects found were a polished soapstone (steatite) bowl rim shard, a plummet, and celt, all of which, like the flint points, had to have been traded into Florida from the north. Neither steatite nor flint is found anywhere on the Floridan Peninsula.

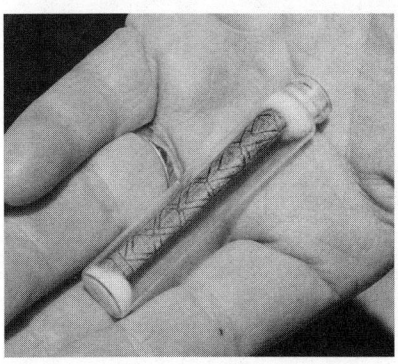

The small vial preserves a length of hollow bird bone. While its use is unknown, the intricately carved pattern attests to the skill and patience of the Tequesta artisan.

Bone, especially deer bone, played an important part in the life of the people of the almost completely formed Everglades area. The site yielded an unusually large number of bone bipoints, socketed points, gouges, awls, pins, even a bone carved bird effigy drilled for decoration as an amulet. Socketed bone points were shaped by cutting small mammal or bird bones off square on the

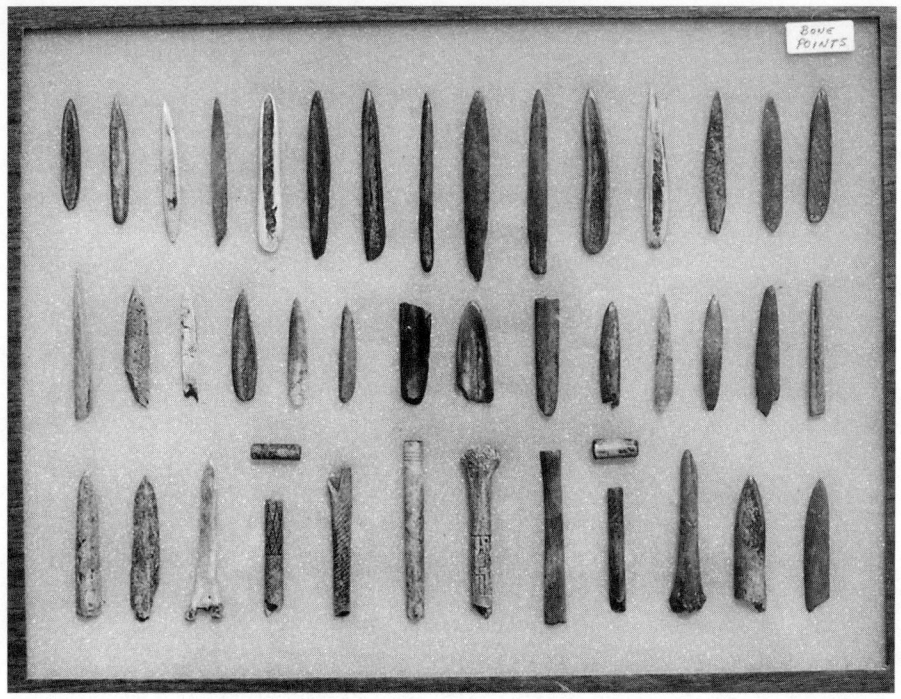

This collection of carved bone awls, points, and daggers was found in a habitation site in the east glades not far from Miami's busy Kendall Drive.

ends. A diagonal cut in the middle produced two sharply pointed tubes, each with a natural opposite end socket into which a shaft could be forced. Pieces of worn pumice from Central American volcanoes, that probably were carried by Gulf currents around to the South Atlantic coast, were used as abrasives for grinding and honing the bone and stone tools and decorative items.

Consistent with many area sites was a hard concreted layer or layers throughout the strata of the mound. In this case, it was a blackish rough material containing bits of food bones, garfish scales and vertebrae, shell, pottery, charcoal, and ashes; it simply is consolidated garbage.

The people interred their dead over a period of thousands of years in a burial mound nearby. A few also had been interred in the site. The clusters of larger bones appear to have been gathered from corpses after long exposure in the open, secondary burials.

An occasional shelter, indicated by a posthole with a flat stone in the bottom, was constructed by the earliest inhabitants. Hunting and gathering dominated the way of life. Later, the introduction of pottery seems to have been the only major cultural change. The flint and soapstone artifacts and the unusual variety of pottery types most likely were trade items that testify to the wanderings of the people over a very long period of time.

There seems to be little doubt that earliest human habitation of the Markham Park Site No. 2 mound began soon after the hammock was high enough above the surrounding waters to keep people dry and their campfires burning. However, investigating archaeologists deduced that the mound apparently was not a permanent living site. More than likely it was only one of a chain of campsites found throughout the Everglades that were utilized between destinations. These are easily located today on elongated tear-drop shaped tree islands, with the campsite located on the higher elevation on the upstream northern end.

Throughout time, people living in low coastal areas or areas prone to flooding have sought to live on higher ground. Where elevations were low and no higher ground could be found, they raised the land with the most durable and abundant materials at hand: earth and shells either from freshwater streams and lakes or coastal marshes and lagoons. This seems to be particularly true for the distal end of the Floridan Peninsula. By about 6,000 to 5,000 B.C., the utilization of shellfish as a source of nutrition provided people with an inexhaustible supply, as well as a seemingly indestructible source of building materials (of course, mound construction reflected availability of local materials, i.e.; shells, sand, dirt, rocks, or any combination thereof).

Until the early part of this century, the southern end of Lake Okeechobee was free to flood during times of water levels that parallel those of today. According to John Fritchie's *Everglades Journal*:

> Old indians, hunters, settlers, and soldiers [of historic times] told how water south of the lake would rise four feet overnight and six to eight feet in a day or two. If a north wind blew for two or three days it blew the water of the lake into the Everglades and the water rose forty miles below the lake four or five feet or more ... and this was just water from rain several miles north of the slough.

Why humans built mounds in the area is obvious. Shell mounds that were used for habitation are known as middens, and often contain artifacts that amounted to the people's trash at the time of their disposal, as noted at the Markham Park Site. An hypothetical Archaic Period mound that was used into historic times in cross section might reveal strata containing at the bottom wooden artifacts, bone implements and points, atlatl parts, and other remains of the oldest Archaic culture. Above that, layer after layer, spanning thousands of years, fiber-tempered pottery is introduced, decoration is added and then becomes more complex, styles and

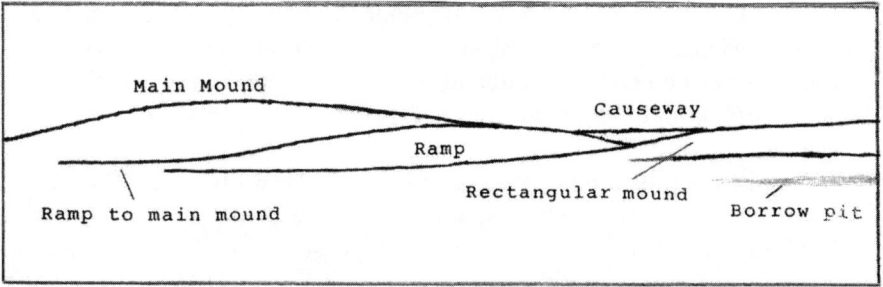

Top: Madden's Hammock is the Miami Lakes complex comprised of three major mounds connected by elevated causeways and reached from today's ground level by sloping ramps. *Bottom:* The author stands atop the highest; the other two are elongated rectangles stretching several hundred yards west. A single small mound is about 50 yards east. East of it are small lakes and wetlands, vestiges of the glades that once flooded the area. Although modern development threatens just across the street, the complex should survive under protective laws to remind us of the ancient Tequestas who laboriously raised themselves above surrounding waters.

tempering materials change, tools and points are made of shell or stone, as are knives and scrapers. This layering of accumulated materials in strata containing specific types of artifacts delineates time frames within the archaeologist's chronological sequence.

Extensive shell middens were formed throughout southwestern

Florida from 3,000 to 2,000 B.C.[4] Horseshoe-shaped ridges, similar to those found in Georgia and South Carolina, have been located on Florida's southern Gulf coast on Horr's Island and at Bonita Springs, another on the east coast near Jupiter. Other sites of the period are reported within the eastern Everglades.

Shell mounds are not peculiar to the southern tip of Florida. They occur along the entire Gulf coast northward to the panhandle and beyond. Shell mounds are found on the upper east coast of Florida as well, and at scattered intervals along the eastern seaboard as far as Maine. In fact, it might be noted that everywhere on earth principal builders of mounds have been maritime people or at least people living along great rivers. This applies especially in countries famed for the size and extent of their prehistoric shell heaps, where the people sought to elevate themselves and their dead above water. In the latter case, burials sometimes were made in the floor of the original ground surface, then on the original ground level, then in the mound in layer upon layer of newly added material. Some were buried in the flesh in extended or in flexed position, while others were defleshed and their skeleton was interred in anatomically correct order and position, or more simply as bundles of bones. This practice convinced early grave robbers and a few archaeologists to jump to the false conclusion that these people had been cannibals. Occasionally, decapitated skeletons or just the skulls were buried. Some remains were cremated.

By building special mounds upon the shell islands for foundations for temples, storehouses, public buildings, and dwellings for religious leaders and chiefs, they were more protected from the high water of hurricanes, tidal waves, or floods that might sweep the area. Building of the mounds also placed the people near their chief sources of sustenance.

Not every archaeological investigation, professional or otherwise, yields hypotheses regarding the past that withstand scrutiny. An example involves laymen excavating in the field in the 1800s who erroneously deduced that earliest inhabitants of the lower Floridan Peninsula came across from the Caribbean, perhaps the Antilles, or South or Central America, rather than the generally accepted Bering Strait, North American migration pattern previously outlined. The other Pacific and Atlantic routes previously mentioned were not found in the literature of that time. To these laymen, artifacts they uncovered appeared to provide convincing clues for such a northward migration. They felt that earliest Floridians used throwing sticks that were thought to be South or Central

---

[4]*The dynasty of the pharaohs in Egypt lasts from about 2200 to 525 B.C. and earliest Eyptian mummies date from the beginning of this period.*

American in type, and repeatedly displayed decorative details found in carvings in Yucatan on the Central American peninsula. Ancients of both areas had the common use of seashells for ear buttons and labrets (lip discs), tools, and other products. Nearly all working parts of tools and weapons came from shells or teeth of creatures of the sea, which only attests to the ingenuity of early man and his ability to meet the needs of life with the resources at hand. Some of the fiber-tempered pottery shards found in the Southeast were similar to pottery discovered on Colombia's Caribbean coast. The same types of pottery dating back 4,000 years ago to about 2000 B.C.[5] are found as far north as Georgia and South Carolina, possibly having reached there by trade or transfer of manufacturing knowledge from one traveler to another. That, they thought (erroneously), strengthened their hypothesis regarding earliest inhabitants' travel by sea on northeasterly currents from South and Central America to coastal Florida. Most other Indians of the southeast did not begin to make pottery until about 200 B.C. Nonetheless, all hypotheses regarding Caribbean origin have proven impossible to confirm. To date, study of these cultural materials has failed to prove any Caribbean, Central or South American origin.

Back in the late 1800s, a Captain Collier, digging garden dirt from a site that eventually would prove an archaeological treasure, uncovered a polished whelk ladle and cord-like material which he felt were of ancient origin. Collier and client Charles Wilkins, a tarpon-fishing winter visitor, shared their fascinating finds with a Colonel Durnford, digging up conch tools, ladles, wooden cups, and a carved wood animal figurehead, all of which had been preserved in the anaerobic mangrove muck. Several weeks thereafter, Durnford took their finds to the curator at the museum of the University of Pennsylvania and ethnologist Frank Hamilton Cushing. Ethnology is the anthropological study of socio-economic systems and cultural heritage, especially of cultural origins and factors influencing cultural growth and change, in technologically primitive societies.

Recognizing the significance of the artifacts, Cushing immediately traveled to Punta Gorda on Florida's southwest coast. He sailed out to explore the promising islands within Pine Island Sound, San Carlos Bay, and as far south along the Gulf Coast to Marco Island. Of the latter he wrote "never in my life, despite the sufferings this labor involved, was I so fascinated with or interested in anything so much, as in the finds thus daily revealed."

---

[5]*Stonehenge is constructed as a religious center in England during the Bronze Age. Babylonians use highly developed geometry for astronomical observations.*

The initial findings at Key Marco prompted more detailed work by Cushing. He would recover an unequaled collection of Calusa artifacts, present an initial report on the sites in November of 1896 to the American Philosophical Society, and earn the reputation for which he is known to this day. He wrote, albeit mistakenly and rather lyrically, that he believed he was

> not alone in thus having found a decided correspondence between the arts of ancient Floridians and other southern Indians and those of the ancient Yucatan. Numerous other observers have noted unmistakable similarities between the arts of Yucatan and Mexico, and those of the mound builders of the Gulf States. It has been held that these arts traveled overland in some way along the far-reaching western and northern Gulf shores from south northward. However, arts and especially ceremonial and decorative art forms do not readily travel from one tribe to another, unless both peoples are in a very similar grade of culture development or share a common environment in which these arts are natural and at home. Moreover, it is to be reflected that not only arts, but also peoples (in sufficient numbers to impress their culture or arts upon others) travel very slowly by land impeded in their course by tribe after tribe, danger after danger. Both arts and peoples travel with utmost facility by sea. Probable traces of Caribbean art have been found.... Such traces of Antillian art as are found in the region of the Key dwellers and farther north on the western, or Gulf Coast, seems to be rather more ancient than the date of the Caribbean occupation even of the West Indies themselves, that is, they seem to be far more Arawak than Caribbean, and this again coincides with the idea of a very far southern origin [in the beginning] of these peoples of the Keys.

Unfortunately, Cushing died within two years of his exciting discoveries, but his initial findings and papers are significant, providing insight into aboriginal life as he came to understand it. Thus, he is quoted throughout the remainder of the section on the Caloosahatchee and Ten Thousand Islands Areas.

Arawak Indians were thought to have migrated from the South American Orinoco region to the east coast of Venezuela and thence out to the Antilles. A few years ago investigative work by a linguist claims to have demonstrated that the major language spoken in Florida in historic times was not really related to other North American Indian languages, but was related to one spoken in the Orinoco region of South America. That linguist also apparently was hypothesizing about the possibility of a northward migration.

In vain, former Florida State Archaeologist Vernon Lamme strived for decades to substantiate the same hypothesis. He writes:

The Archaic Period

> For the past forty years I have believed that some day it will be proven that the ancient Maya Indians of Yucatan, Guatemala, and Mexico whose civilization seems to have sprung up in full bloom with no artifacts found there denoting a primitive existence, originated in the peninsula of Florida, many centuries before Ponce de Leon.

His theory was based on his find of a three-inch-tall carved stone statue of a kneeling man wearing what appeared to resemble an early type of leather football helmet. Some years later, another archaeologist discovered immense stone heads weighing many tons of identical facial features wearing the same type of helmet in the jungles of Yucatan. A few other artifacts of parallel type were found in both areas; those from Florida were thought to predate those attributed to the Maya. Lamme never was able to substantiate his theory.

As strong as arguments seemed to be regarding a connection between Yucatan or the Caribbean and the southeast, it is important to understand that these hypotheses regarding the very earliest migration to the Floridan Peninsula remain unfounded due to lack of proof. They certainly not only lack support, but are considered nonsense by most modern professional archaeologists.

Irving Eyster, retired former Miami–Dade County archaeologist, has written:

> Every two or three years someone finds potsherds, whistles, or figurines that definitely are Mayan, but so far, all of these were from historic Indian mounds or from the beaches where they were washed up by hurricanes. The Indians of the historic period were extremely good divers, and were often employed by the Spanish to help salvage wrecks. So it is understandable that they may have done some salvaging on their own. Most of the Central American artifacts found here were from the 1733 wrecks. I have been looking for a prehistoric mound or site that has artifacts of the Central American culture for the past twenty years, but have never found anything I could prove were trade items or that were brought by immigrants of this great civilization.

While early human entry onto the Floridan Peninsula from the south millennia ago remains unproven, certainly there was nomadic expansion much later. Travel is known to have occurred from the Antilles and/or Mesoamerica (central Mexico south to the Pacific coast of El Salvador). By that time, huge canoes, some locked together in catamaran fashion, were more than capable of navigating the Caribbean and Gulf. Surely, by 500 years ago, aboriginals of Hispaniola and Cuba would have sought escape from bondage, torture, and death at the hands of Europeans. Could

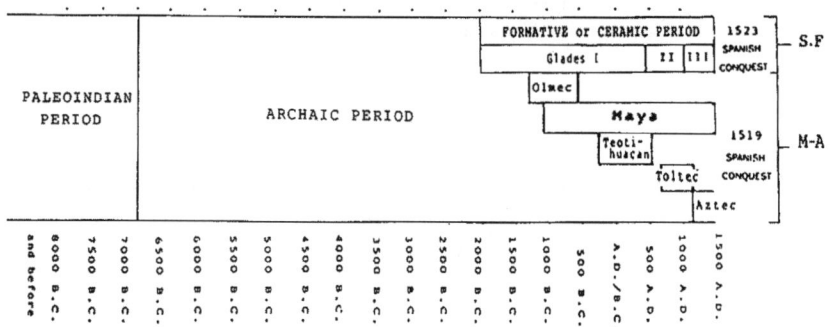

The time ladder compares the more widely known cultures of Mesoamerica ("M=A") with Paleoindian, Archaic, and Formative periods of the Floridan Peninsula ("S.F."=South Florda).

they have made it? Why not? With favoring winds and currents, people make it on little more than the crudest of rafts to this day.

Isolation of approximately the southern third of the Floridan Peninsula from the northern two-thirds by virtue of the vastness of the wetlands, the glades, surely would have minimized, but certainly did not negate diffusion of cultures, as indicated by the diversity of artifacts found in sites such as Markham Park.

Consequently, throughout the enormous span (in human terms) of the thousands of years of the Paleoindian Period, plus some 5,000 years of the Archaic Period, human utilization of the evolving Floridan Peninsula steadily became more intense due to the aboriginals' increase in numbers and their inventiveness. Effectiveness of the atlatl was improved by addition of a sliding weight (bannerstone) to the throwing device. The utilitarian advantages of fire-hardened pottery were being discovered. Mobility increased dramatically in this watery realm through development of the dugout canoe. With its advent, aboriginals learned more fully to utilize the bounty of the swamps, lakes rivers, and sea.

By the Late Archaic,[6] the diet of the people of the Floridan Peninsula was predominantly from freshwater and marine sources. Fishing, including taking of sea turtles, probably accounted for all but a small percentage of the diet of those living at coastal sites. Hunting of mammals and reptiles provided a rather small portion of the diet, while birds seem to have been a scarce contributor.

[6]*Moses receives the Ten Commandments on Mount Sinai. The time of the beginning of the Olmec culture in Mexico and the Minoan Period on Crete. A rock carving in Norway depicts the earliest representation of skiing.*

The diversity of botanicals continued to broaden as the environment evolved to its semi-tropical form. Of course, the difficulty of finding and identifying botanicals in archaeological context very well might distort the picture. Nonetheless, plants obviously were utilized to a more significant degree than the archaeological record currently indicates. Milling stones, such as the set found at Santa Maria, for grinding wild seeds, nuts, and other vegetable and animal matter, confirming the point, are found in the Archaic Period context. One example might be that of acorn preparation for human consumption. Certain species of acorns contain toxins and are bitter to taste. The Archaic people had learned that hulling the acorns, then grinding them into flour and leaching it in water removed the toxin and bitter taste and rendered a nutritious food.

Despite the developing number and size of shell mounds, the meat of the mollusks themselves is not credited with contributing more than a minor dietary percentage. As nutrition improved, it is reasonable to assume that the general state of health more than likely improved as well.

Comparison of the known size and number of Paleoindian sites with the many larger and far greater number of Archaic sites that contain higher numbers of artifacts indicates that the Archaic Period population had expanded considerably and there was an exchange of trade goods. Recovered man-made artifacts in the sites display a bridge of the gap between hunters and sedentary people, charting a cultural evolution that occurred gradually over many thousands of years. Families and/or social, political, or warring groups mixed gene pools, and with that, physical characteristics probably altered somewhat as well. The diminutive, wiry, five-foot four-inch, 110 pound individual whose physical features revealed Mongoloid ancestry, in spite of being a composite of racial strains, in 10,000 years had developed into a human of only slightly taller, heavier, and perhaps more robust stature.[7] Probably average life spanned somewhat longer than the Florida Paleoindian's 30-odd years.

---

[7] *By 800* B.C. *(near the end of the Archaic Period) the Greek Homer had written both the* Iliad *and the* Odyssey. *The first recorded Olympic Games had been held in 776* B.C. *Within another hundred years, water clocks in Assyria kept track of time, and work on the Acropolis in Athens had begun.*

4

The Formative or Ceramic Period: 2,000 B.C.–A.D. 1513

Prior to the development of radiocarbon (C14) dating technology, the advent of ceramics in the Old World as far back as 12,000 years ago provided archaeologists with chronological markers of reasonably accurate dimension. Specific stone point types, such as the previously mentioned Clovis Point, for example, also provided chronological markers. By comparison to other North American geographic areas, use of stone on the Florida Peninsula was extremely limited and, so failed to provide the necessary measure. Barbara Purdy (1996) writes:

> Pottery objects became culturally and chronologically specific, so that today archaeologists can recognize the spatial and temporal origins of literally hundreds of designs and forms. This knowledge is especially valuable in the absence of stratigraphic control and when no absolute dating method can be utilized.

The oldest pottery found so far in Florida dates back 4,000 years. Consequently, utilizing criteria of design, decoration, and temper, archaeologists have been able to create reasonable time divisions within the Formative or Ceramic Periods of the Floridan Peninsula.

Earliest makers of pottery discovered that inclusion of Spanish moss, yucca, or other vegetable fibers in the clay, or paste as potters call it, held the material together during shaping and subsequent firing. Heat of fire destroyed the organic fibers, leaving unmistakable fine hollow tracks throughout the pottery. Without the fiber temper the clay often would crack, or the ceramic piece would become distorted. These telltale fiber tracks date this fiber tempering process back to its beginning on the Floridan Peninsula to about 2,000 B.C. Lack of pottery prior to this time may be a consequence of the people's nomadic lifestyle; it simply was too heavy

The Formative or Ceramic Period 75

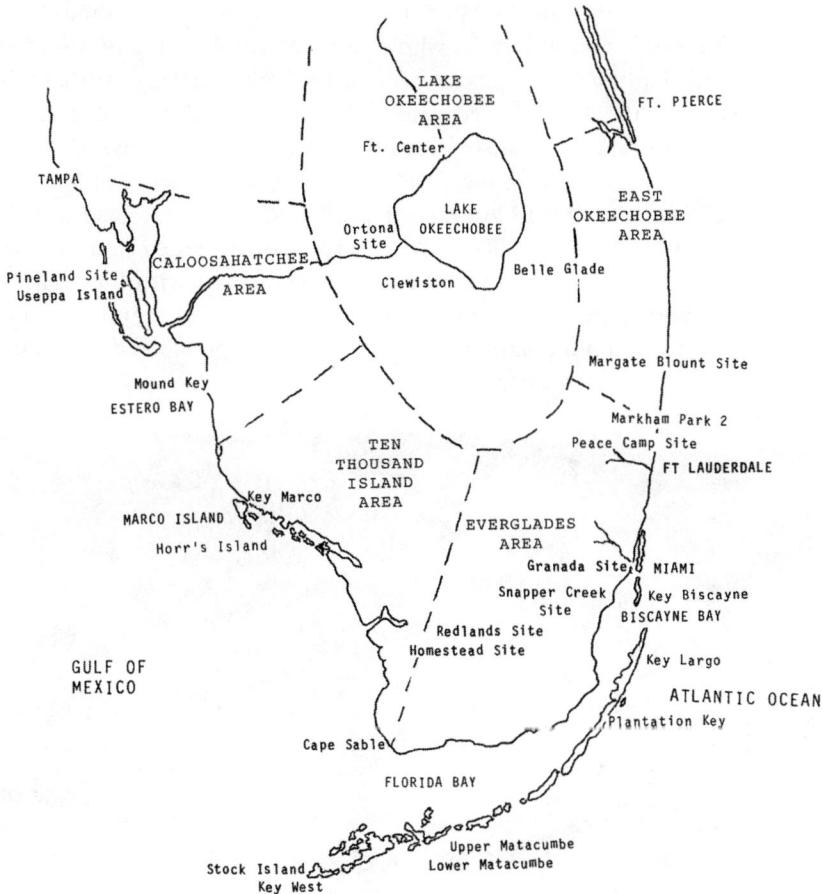

The Floridan Peninsula of 500 B.C.–1,500 shoreline approximates that of today. The Keys are reduced to a chain of islands and coastal bays supporting rich estuarine mangrove environments. Lake Okeechobee has long since filled and supports the unique glades environment to the south. Archaeologists divide the map into Cultural Areas. (Not to scale.)

and cumbersome to carry around. Its introduction generally coincides with adoption of a more sedentary way of life.

In due course, potters blended both fibers and sand into the paste, and this advance became known among archaeologists as semi-fiber tempered ware, and is construed by some to delineate the next time frame, but by others to fall within the same type as the former from 2,000 to 500 B.C.[1]

---

[1] King Nebuchadnezzar builds the legendary "Hanging Gardens" of Babylon.

The durable fiber and/or later crushed shell or pottery, or sand-tempered ware rarely is found in its whole form. Most often the bottom has been punched out or the entire piece fractured. Some writers state that this was a symbolic practice referred to as "killing," in which the spirit of the ware was encouraged to transfer from the old piece to the new.

The time frame within the Formative or Ceramic Period of 500 B.C. to A.D. 1700 for the south Florida area is divided into three Glades Periods. Glades I ranged from 500 B.C. to A.D. 500, during which time pottery was hand-molded, mostly undecorated, and tempered with fiber. However, during the latter part of the Glades I Period, crude ceramic decorations (incising and punctuating, linear carving, and punched depressions respectively) were introduced.

Tiny filamentous holes throughout the paste of fiber-tempered pottery reveal a manufacture date possibly as far back as 2,000 B.C. The clay was mixed with plant fibers for stability. The firing process burned away the organic fibers leaving their traces behind.

During the Glades II Period, A.D. 750 to 1200, pottery types were incised with a sharp instrument prior to firing, and this decorative type continued with numerous variations and pinched rim treatments (here again, thought by some to suggest Antillian influence). Wooden paddles wrapped with fiber cord were utilized to impress patterns on the ceramic's paste surface. Temper of sand and, occasionally, crushed shell or pottery became more common.

The Glades III Period, A.D. 1200 to 1700, obviously is inclusive of the time of European contact with native American Floridians at about A.D. 1513. Pottery of the distal end of the peninsula of this period became quite distinctive in shape to the trained eye. Here, too, tempering materials included sand, crushed limestone, crushed shell, and/or bits of old pottery.

Potters started forming ceramics by coiling, rather than hand-molding of the clay. Additionally, a trade network brought a broad variety of non-native materials into the area in the form of beads, points, implements, and the like.

Top: Pottery tempered pottery and (*bottom*) sand and shell tempered pottery.

Some in the field believe there is sound evidence, disputed for earlier times, that the Arawak Indians, who had originated in the Orinoco and migrated to Cuba, now reached south Florida. They feel that the Arawaks' Antillian influence upon the area's people at this later time is evident. For instance, Carr hypothesizes that Arawak arrival in South Florida from the Bahamas or Cuba by A.D. 1200 to 1500 may have influenced the use of the root of coontie (*Zamia integrifolia*) by locals as a source of starch, as it is similar to that produced from the root of Caribbean manioc (also known as cassava) for bread and, later, tapioca.

Changes in ceramics in the Glades sequence are really not that dramatic, and many more variations and time divisions within each period than are described here are utilized by archaeologists. Those presented are sufficient as chronological markers for the purposes of this work.

CERAMIC SEQUENCE IN FLORIDA

Periods	Dates	Distinguishing Ceramics
Archaic Period	before 2,000 B.C.	pre-ceramic in Florida
Formative or Ceramic Period		Glades Plain, fiber tempered
Glades I	to 500 A.D.	Glades Plain, semi-fiber tempered, sand tempered
Glades I Late	to 750 A.D.	Fort Drum Punctated
Glades II A	to 900 A.D.	Key Largo Incised Miami Incised Opa Locka Incised
Glades II B	to 1,000 A.D.	Key Largo Incised Matacumbe Incised
Glades II C	to 1,200 A.D.	No decorated ceramics
Glades III A	to 1,400 A.D.	Surfside Incised
Glades III B	to 1,500 A.D.	Glades Tooled Rim
Glades III C	to 1,700 A.D.	Glades Tooled Rim

Pottery predating 2,000 B.C. has not been found to date on the Floridan Peninsula. Thus, time prior to that date is considered preceramic. Many additional designs and variants thereof are listed in the ceramic sequence. The above are presented solely as examples.

The Formative (Ceramic) Period ceramic sequence is presented in simplified form solely to indicate how archaeologists utilize pottery as a dating agent. Pottery predating 2,000 B.C. has not been found on the peninsula; therefore, time prior to that date in Florida is considered pre-ceramic. Many additional designs and styles are named within the ceramic sequence, and the number is further compounded by numerous variants. Those displayed serve only as examples. (Pottery shard sketches: Irving Eyster.)

For additional clarification, type names are assigned to various ceramics that reflect their type site, as was the case for the previously described Clovis Point, the type site for it being Clovis, New Mexico. Consequently, with these south Florida pottery types there are names such as Glades Plain, Miami Incised, and Fort Drum Punctate to name but a very few, each being associated with one, two, or all of the Glades Periods. While these, plus their variations, provide a needed tool for archaeologists, they can prove confusing and are touched upon only briefly here and in the accompanying illustrations.

It is of interest to note that some of the earliest records of fiber cord and woven or plaited fabrics are found as impressions left on pottery. Generally, such material of rather delicate organic origin decomposes rapidly, unless submerged under anaerobic pond bottom muck (or, in other geographic areas, in extremely dry environments, or preserved frozen). In some instances, the still wet unfired pottery was placed on a woven fabric or mat surface, and the reverse of the weave became indelibly imprinted in the bottom of the vessel. So, while the fiber and fabric has been lost for millennia, its impression in the ceramic has remained as an almost indestructible archaeological record of the weaver's art. Additionally, residues in the pots, even minute traces of materials stored within the pores of the ceramics, can serve as a source of dietary and other information.

The southern third of Florida has been divided into five loosely defined archaeological cultural areas. Apparently there must be a great deal of area overlap in every direction due to warfare, trade, resettlement, and other factors that virtually eliminate the possibility of fixed geographic boundaries. Each area has been roughly delineated by certain general cultural characteristics that were adapted to local ecological conditions and shared by the human population. The five South Florida cultural areas are entitled by Carr and Beriault as East Okeechobee, Lake Okeechobee, Caloosahatchee, Ten Thousand Islands, and Everglades.

East Okeechobee is the most poorly defined, but roughly includes the area from the Broward–Palm Beach County line north to the St. Lucie Inlet. The Okeechobee area surrounds the huge lake after which it is named, and includes a portion of the Kissimmee River Basin on the northern side of the lake. The Caloosahatchee area occupies the northern part of southwest Florida from about the Cape Haze peninsula and Charlotte Harbor south to Estero Bay. The Ten Thousand Islands area includes the unique island area south of Estero Bay to Cape Sable, eastward into Big Cypress swamp and northeast to the southern end of Lake Okeechobee. The Everglades area encompasses much of southeastern Florida, inclusive of the Keys, and from Cape Sable northeast back to about the Broward–Palm Beach County line.

Ceramics types, tool and weapon types, all artifacts and ecofacts that escaped the ravages of time, basically help define the Indians lifestyle within a geographic locale. This is why the locale is referred to as a "cultural area." The people who lived within that area, with whom the artifacts and ecofacts were associated are, for the first time in Florida prehistory, accepted by anthropologists and archaeologists as having a group, band, or chiefdom name. The names are those as they were understood and transcribed by the earliest Spaniards with whom the natives had verbal discourse. They are the Ais of the East Okeechobee area, the Mayaimi of the Okeechobee area, the Calusa (Caloosa)of the Caloosahatchee and Ten Thousand Island areas, the Tequesta (Tekest or Tekestan) of the Everglades area, and others.

In addition to pottery, other new styles of artifacts and new technological skills made their appearance during the millennia of the Formative stage. The people who possibly were antecedents of Archaic and still older

Paleoindians, or perhaps were later migrants onto the peninsula, might have begun to develop agriculture through cultivation of some plants, including, possibly, the controversial maize or corn. Other plants, such as the wild gourd-like squash (*Cucurbita* spp.), bottle gourds (*Lagenaria siceraria*), wild cotton (*Gossypium hirsutum*),[2] and Spanish bayonet (*Yucca aloifolia*) simply might have been encouraged to grow nearby. Agriculture remains a bone of contention since the Spaniards recorded the people, at least of the southern part of the peninsula, as having none. The archaeological record, thus far, seems to substantiate this 16th century documentation.

Plant materials can be preserved in environments that are too dry or too wet (or frozen) for survival of decay organisms. Archaeobotanists are able to collect samples by screening. Wet materials are screened and then floated in water. The lighter floating organic samples then are separated from the liquid for analysis. Carbonization due to fire also may provide sufficient samples to permit identification. Wood identification by microscopic comparison with modern wood specimens is usually made to the taxonomic level of genus. For identification to the level of species, attendant vegetative and/or reproductive plant structures must be present. Seed analysis is conducted utilizing size, shape, and surface texture as comparative visual features. Hard rind, such as that found on gourds, may survive the millennia, as do nut shells and pieces of the tough cob of maize.

Palynology, plant identification through pollen analysis, provides additional clues for the archaeobotanist. Once treated with strong alkali and acids, washed, and stained, the pollen exhibits a distinctive shape under a strong microscope and is identified as to genus and often to species. These organic and environmental remains, including floral and faunal materials, as well as soils and sediments, the non-artifactual materials, are what are referred to as ecofacts.

Anthropologists tell us that Formative Period communities became more settled and more complex with the evolution of religious and political organization. Habitation sites evolved into islands of accumulated debris and fill (some of which actually were started during the Archaic Period). Coastal shell middens, house mounds, platform mounds possibly for ceremonial use, and burial mounds often were protected from offshore elements by artificial breakwaters. From within the lagoon or bay created by the breakwater, canals led to habitation sites and areas of supposed

---

[2]*Wild cotton is an ancestor of today's commercial cotton. It has become rare because of a 1930s government eradication program designed to protect domestic cotton from genetic cross-contamination with its wild ancestor.*

religious significance. Walks, ramps, even sunken paths led to and from these major fixtures. Many of the man-made works were of shell, and earth, or sand. Some bore a facade of a single species of shell, the narrow end (siphonal canal) of each having been thrust into the fill. This engineered a more stable, aesthetically pleasing structure. More inland habitation sites, while high and dry today, bear similar design as ecological adaptations to what formerly were wetlands and flood-prone geography. All such accomplishments had to have taken place over generations of time, by large numbers of people, under skillful leadership of a succession of chiefs.

EAST OKEECHOBEE CULTURAL AREA

Margate-Blount Site

Broward County's Margate-Blount site presents a fine example of an East Okeechobee cultural area site. It is located on the eastern edge of the glades about 12 miles west of today's Atlantic coast. This was a long term habitation area with an adjacent burial mound, the latter originally rising to a six foot elevation above the surrounding terrain. Chronology of the site indicates that it, too, was utilized throughout the entire latter half of the Formative Period, 500 B.C.–A.D.1500.[3]

Of particular interest is that excavation revealed what might have been remains of a wooden structure, beneath which alternating layers of ash and soil indicated that a charnel house had existed on the mound. When property owner Bruce Blount described his early recollections of the mound, he commented that in the 1940s it was surmounted by a ten foot long, by six foot wide, by four and one-half foot high log enclosure containing skeletal remains "stacked like cordwood with no earth in between." Most of that surface structure and its contents were later lost. The burial mound was damaged during a bulldozing operation to level and expand the area for sod farming. When the blade bit into the mound, it scattered many human bones.

Between the mound and the habitation area, there was a lower wetland area into which a modern drainage canal had been dredged by the Corps of Engineers. Burials were dredged along with the muck and deposited entirely out of context on the canal embankment. However, a proper scientific investigation yielded primary and secondary burials that

---

[3]*The last queen of Egypt, Cleopatra, died in 31 B.C. and Caesar was murdered in 44 B.C. Jesus Christ was born at Bethlehem and was crucified some 30 years later.*

totaled 49 individuals, not inclusive of those displaced by the heavy equipment. The collection is one of the largest prehistoric skeletal samples from southeastern Florida.

Primary human burials consist of interment of the whole body in an extended or flexed position. Secondary burials were accomplished by defleshing the skeleton. Thereafter, the bones often were stored in a charnel house, such as the one described by Mr. Blount, for burial at some later time. Upon interment the bones were arranged in a bundle, or laid out in correct anatomical order, or even arranged in some fanciful display meaningful only to the people and their spirits. At Margate-Blount one skeleton was covered with a wooden slab, beneath which, when it was removed, was found a child's secondary burial above the skeleton's feet. The child's vertebrae and ribs had been arranged in a circle around its skull, and its other bones had been placed beneath and beside it. Other wooden artifacts associated with the burials were a five foot, two inch long wooden paddle, possibly of cypress, and a heavy pestle carved from unidentified wood.

Some 4,000 pottery shards were recovered from levels throughout the site that were representative of all of the Glades Periods within the Formative Period. A few from the uppermost level proved to be of non-local types, indicating trade with Indians from the vicinity of today's St. Augustine.

Shell artifacts included Queen Conch celts, scrapers, a bead, whelk shell tools and vessels, a Sunray Venus knife, columella tools, shell gravers, and drilled scallop shells. Bone artifacts included drilled shark vertebrae and teeth, socketed points and bipoints, gouges, knives, beads, a drilled alligator tooth and drilled human tooth, a stingray spine, a sawfish bill, and a drilled turtle plastron that might have served as a gorget.

The lithic assemblage included locally made pendants of limestone and coral, and several of non-local stone, including a grooved polished black pendant possibly carved to represent the head of a sea turtle.

The earliest, lowest levels of the site did yield quantities of alligator and deer remains. However, the later subsistence behavior of the people of Margate-Blount was primarily oriented toward freshwater gathering, with limited hunting. Shallower later levels indicate that the people more and more exploited smaller fish and reptiles, as well as coastal marine resources, as indicated by the shark teeth and vertebrae, stingray spine, and sawfish bill. It is of interest to note that points removed from along either side of a sawfish bill often were utilized by early people as fishing gorges. Tied in the middle to a woven fiber line and baited, the gorge lodged in the throat of the fish unlucky enough to swallow the mass, and was easily taken.

Skeletal biology of the human population revealed that the 49 individuals represented a sex ratio of 20 males to 22 females, the balance being indeterminate. Age distribution was three individuals two years or younger, five between ten and 20, 18 in their 20s, 13 in their 30s, seven in their 40s, two in their 50s, and one over 50. The ratio of males to females in the age groups indicate that the males outlived the females in this sample population. The estimated adult stature averaged about five feet six inches for the males, five feet four for the females.

Dental problems of antemortem tooth loss seemed more common among females. Only one female had a cavity. Abscesses were more prevalent among the males, but none displayed any evidence of cavities. Tartar deposits on all were minimal and dental wear was slight to moderate, certainly less than that observed for earlier people of the Archaic and Paleoindian Periods.

Bone pathology revealed that the Margate-Blount people suffered some inflammatory bone lesions, arthritis, and one individual clearly displayed a healed fracture of the right ulna. While this group does not represent a statistically reliable data base, comparison with descriptions of earlier people does appear to demonstrate that, on average, they seem to have gained an inch or two in height and, perhaps, a number of years of longevity.

Investigation of the Margate-Blount site during the three year period of 1959–1961 resulted in the founding of the Broward County Archaeological Society.

LAKE OKEECHOBEE CULTURAL AREA

Fort Center

The Fort Center site, located on the north side of Lake Okeechobee, is within an area about one mile long by one and a half miles wide. It was in existence as a human habitation site for over 1,000 years through the Formative Period, having first been occupied as early as 2,000 years ago.

The area around the site frequently is flooded, and construction of earth structures and resultant variable settlement patterns were adaptations to this particular fluctuating environment. Earliest inhabitants, perhaps one or two families at a time, lived along the river that drained into the lake on small house mounds well out into the river's meandering belt. This placed them near the source of fish, turtles, and shellfish upon which they relied for food they prepared in simple, fiber-tempered ceramic bowls.

For many years archaeologists thought the Fort Center site was unique because of its great circular mounds and ditches. Three were sequentially constructed and utilized: the first two were about 300 feet in diameter, and the third was 1,200 feet in diameter. William H. Sears, Professor Emeritus of Anthropology, Florida Atlantic University, hypothesized that the circles were utilized as drained agriculture plots, after similar plots found in South America. Maize pollen was in evidence in samples found in the circle interior, from midden deposits along the river bank, and at the ends of the great circle. There was strong evidence of shell burning to produce lime. Sears felt the lime was used to mix with water, in which stored dry maize or corn would be soaked to make it palatable as hominy. No pollen or other corn material was found in numerous human coprolites, nor was there evidence of cobs from the site. Questions regarding maize cultivation are raised in the literature time and again. Some authors believe the presence of pollen is a positive indication of maize cultivation. Others are equally adamant that prehistoric south Floridians are considered non-agricultural, as documented by early Spanish explorers. The presence of pollen proves nothing and both cob and kernel remains have eluded investigators to date. Museum specimens of cobs from that time are only one and one-half to two inches long. It is reasonable to consider that cob and all were consumed, leaving little behind in the way of identifiable residue. Equally arguable is that at least kernels or portions thereof would be found in coprolites. Domestication and hybridization have increased size and productivity of the original native grass discovered in Mexican caves dating from 7,200 to 5,400 B.P., but the basic characteristics of corn have changed little in all those years. Consequently, for lack of further substantive evidence, the hypothesis regarding corn or maize remains a subject of controversy to this day.

Coprolites are considered to provide data on a single meal, not on an individual's entire diet. They may contain bone and shell fragments, seeds, plant fibers, hair, charcoal, remains of birds, fish and insects, algae, fungi spores, and parasites. Pollen, when present, not only indicates presence of the plants consumed or in the area, but the season when the pollen had been produced.

Other Big Circle Mound complexes are known south of Lake Okeechobee, southeast of the town of Clewiston. Carr has discovered evidence of at least seven additional similar circular mounds. He lists two near Fisheating Creek, one north of the lake in Glades County, and another south of the lake in Hendry County. To the south in Miami–Dade County is Dade Circle, about 585 feet in diameter, presently under fill within the city of Miami, and Miami Circle, 195 feet in diameter under the streets of the south bank of the Miami River (not to be confused with the chiseled-in-rock circle feature at the Brickell Point Site). Nothing in these sites was found to substantiate Sears' idea regarding agriculture. Life at the Fort Center site continued for several hundred years with little change, other than the evolution of pottery to sand tempered plain ware.[4]

About 1,800 years ago, A.D. 200, the focus of Fort Center activities did change. The site became a ceremonial center. A complex was constructed consisting of a hand-dug charnel pond with a wooden platform, and a small brown earthen platform mound containing a bathtub-shaped pit for preparation of human bodies prior to their bones being stacked in bundles on the platform. The wood platform was supported by timbers and extended over the pond. The upper end of the large vertical posts had been carved into stylized birds, cats, bears, and other animals. There also was a habitation mound on a low earth wall that surrounded the complex, and was attached to the mound at both ends. It is estimated that several families lived on the separate mound, what is referred to as the brown mound, and two to four families lived on the habitation mound.

The diet of these people was revealed in the tremendous quantities of bones of turtles, fish, deer, turkey and other fauna. Sears, in 1982, found that odds and ends of broken human bone had definitely been deposited with animal bone, and discarded in the same way. He hypothesized that "cannibalism seems probable."

Pottery found included trade-ware specimens from more distant areas, indicating that the ceramics might well have been included in ceremonial functions. Shell and shark's teeth tools, as well as worn and polished chert points were uncovered in this same habitation mound. The suggestion is that they served primarily for carving the wooden figures that adorned the top of the support posts of the charnel platform. The brown mound contained no such artifacts, but did contain large whelk and conch shells, scraps of human bone, skulls, skull caps, large quantities of human teeth, conch shell dippers, and bird bone tubes, all possibly indicative of human ceremonial function.

The pond deliberately had been excavated below the water table (only two feet or so beneath the ground surface). The elaborate charnel pond platform burned at some point in time, and the entire platform, plus the remains of 300 people representing a population norm of ages and sexes,

---

[4]As the people of Lake Mayaimi struggled to survive, great men of the Old World who were to leave their mark forever on civilization lived their lives. The Greek Herodotus (father of history), 485–424 B.C.; the great Lord Buddha, 480–550 B.C.; Confucius (philosopher), 479–551 B.C.; and the Athenian Socrates (philosopher), 470–399 B.C.

collapsed into the pond. Thereafter, human remains, possibly those retrieved from the water, were interred in a mound that was being deposited over the brown dirt mound. The fill continued for a number of centuries until it had been elevated to a 25 foot height and greatly expanded diameter. Eventually, this burial mound contained 150 bodies (as measured by the number of right ear mastoid bones therein). Remains of an equal number of bodies that had been on the platform were preserved within the muck of the pond. It would seem that Fort Center had functioned ceremonially for a society that occupied the area around Lake Okeechobee and farther northward along the Kissimmee River.

When Fort Center first was inhabited, the distant sea level was two or three feet lower than it is at present. By 2,050 years ago,[5] rising seas had persisted for about 500 years, by which time they were higher than they are today.

Thereafter, a period known as the Scandic climatic episode began, a xeric time of colder temperature that lasted another 500 years. Environmental changes more than likely were the major causative factors in lifestyle changes at this site, as well as others throughout the lower Floridan Peninsula. The relationship between the natural environment and culture is obvious. It was during this Scandic episode that ceremonial use at the site diminished, and, as indicated by the comparatively sparse quantity of artifacts left behind, families endured a more mundane existence.

The time of habitation of Fort Center following its last use as a ceremonial site is the time of the beginning of construction of a system of house mounds with interesting attached linear earthworks, the purpose of which is still subject to question. House mounds were small and built in the open savanna along with their linear earthworks. A few burials were found in the top of one of the mounds, and artifacts interred with them apparently were symbols of rank, such as sand–tempered ceramic pipes, and quartz crystal and igneous rock plummets. The latter had to have been imported as finished items from as far north as Georgia or Alabama, as no igneous rock debitage was found at the site. The people continued to use sand-tempered plain, and a greater amount of Belle Glade Plain pottery. Sears writes that:

> both types, though drab and utilitarian, were technically excellent. Most pottery specimens are hard and ring when struck; surfaces are uniform and rims are neatly finished. The firing technique used to produce hard

---

[5] *The 1,400 mile long Great Wall of China was built. The three language inscription that gave modern man access to the secrets of the Egyptians was engraved on the Rosetta Stone.*

pottery also produced light gray, even nearly white or light tan inner and outer surfaces. A series of new rim forms became popular, particularly expanded flat and comma shaped varieties.

Fort Center occupation continued until A.D. 1700, two centuries into the historic period. Hernando D'Escalante Fontaneda, a captive among the Indians from a child of 13 until 30 years of age, between the late 1540s and mid 1560s, spent much of the time in the territory of Carlos, "a province of Indians which in their language signifies a fierce people." He stated that no one knew the country as well as he. During his travels he learned four of the native languages and served as translator between Indians and Spaniards who fell into their hands as a result of their treasure ships returning from Peru having wrecked along the coast. He is felt to have mentioned the site in his 1574 memoirs when he wrote:

> As far as a town they call Guacata, on the Lake of Mayaimi, which is called Mayaimi because it is very large.... In the midst of the country are many towns, of thirty or forty inhabitants each; and as many more places there are in which people are not so numerous. They have bread of roots which is the common food the greater part of the time and because of the lake, which rises in some seasons so high that the roots cannot be reached in consequence of the water, they are for some time without eating this bread. Fish is plenty and very good. There is another root, like the truffle over here, which is sweet and there are other different roots of many kinds; but when there is hunting, either deer or birds, they prefer to eat meat or fowl. I will also mention that in the rivers of freshwater are infinite quantities of eels, very savory and enormous trout. The eels are nearly the size of a man, thick as the thigh and some of them are smaller. The Indians also eat lagartos [alligators], and snakes and animals like rats which live in the lake, freshwater tortoises, and many more disgusting reptiles which, if we were to continue enumerating, we should never be through.
>
> These Indians occupy a very rocky and a very marshy country. They have no product of mines or things that we have in this part of the world. The men go naked, and the women in a shawl made of a kind of palm-leaf, split and woven. They are subjects of Carlos [Cacique of the west coast Calusas at that time] and pay him tribute of all things I have mentioned, food and roots, the skins of deer, and other articles.

The Lake of Mayaimi of which Fontaneda wrote probably is Lake Okeechobee. The bread of roots to which he referred possibly was made of compti root. Compti, known today as coontie or arrowroot (*Zamia intergrifolia*) is a member of the cycad family and is endemic to pine woods. After grating, a poison contained in the root had to be extracted in water.

Thereafter, the rough material was reduced to a flour and consumed in a variety of ways.

Fontaneda repeatedly mentions that there are no mines and the Indians have no silver or gold, the value of which they only became vaguely aware from the numerous ships of treasure fleets—referred to as "the armada of New Spain"—that wrecked upon their shores and were taken captive. Of another shipwrecked sailor he writes:

> He saw them [the Indians] go and return with great wealth, in bars of silver and gold, and bags of reals, and much clothing. As he was newly captured, or found, and understood not the Indians, I and Juan Rodriguez were the interpreters for this man, and others, as we already knew the language. It was a consolation, though a sad one, for those who were lost after us to find on shore Christian companions who could share their hardships and help them to understand those brutes.

He writes on to explain that many Spaniards were saved when they found their previously captured fellows there who understood the language of the natives. The latter, always "themselves very mean, (for the most so are the people of Florida)," were unable to understand why captives failed to follow their orders, thinking them "rebellious and unwilling to do so. And so they were put to death." One of the Indians' great chiefs asked him

> Escalante, tell us the truth, for you well know that I like you much: When we tell these, your companions, to dance and sing, and do other things, why are they so mean and rebellious that they will not? Or is it that they do not fear death, or will not yield to a people unlike them in their religion? Answer me; and if you do not know the reason, ask it of these newly seized, who for their own fault are captives now, a people whom once we held to be Gods come down from the sky.

Escalante explained, after which the cacique put the captives to a test, using him as translator and another captive to verify what had transpired. Having discovered the truth, the casique ordered that future captives were to remain unharmed, "that one might go to them who should understand their language."

Fontaneda also mentions other captives, "Indians of Cuba and Honduras who were lost while in search of the River Jordan." He explains that the Jordan is a superstition of the Indians of Cuba

> because it is their creed, not because there is such a river. Juan Ponce de Leon, giving heed to the tale of the Indians of Cuba and Santo Domingo, went to Florida in search of the River Jordan, that he might have some

enterprise on foot, or that he might earn greater fame than he already possessed and close his life,-which is the most probable supposition; or, if not for these objects, then he might become young from bathing in such a stream.... The Indians ... to satisfy their tradition, said the Jordan was in Florida; to which I can say, that while I was a captive there, I bathed in many streams, but to my misfortune I never came upon the river. Anciently, many Indians from Cuba entered the ports of the Province of Carlos in search of it; and the father of King Carlos, whose name was Senquene, stopped those persons, and made a settlement of them, descendants of whom remain to this day.

He wrote that the kings and casiques followed the belief, searching for the river that would turn aged men and women back to their youth, to the point where "to this day [they] persist in seeking that water, and are never satisfied ... but to this day youth and age find alike that they are mocked, and many have destroyed themselves. It is cause for merriment, that Juan Ponce de Leon went to Florida to find the River Jordan."

A number of years ago a backhoe operator working just west of Clewiston, on the south edge of that lake, pierced an ancient site while digging an irrigation ditch through a ridge that once separated Lake Okeechobee from the Everglades. Among artifacts archaeologists recovered were pottery shards, a shell tool the shell of which had to have originated in the Gulf of Mexico, and human bone fragments. Clewiston and similar sites, such as the one at nearby Belle Glade, indicate rather long term habitation from 500 B.C. to about A.D. 1500; that is 2,000 years, a very long time. The people of the area surrounding the lake were understood by the Spaniards, who arrived at that latter time, to be called by the same name as the great lake, the Mayaimi.

Ortona Site

The Ortona Site is located just West of Lake Okeechobee, near the Caloosahatchee River on the northwest edge of the Everglades. It is a large complex that first was reported about 1835 by surveyors working in that area. The site has been referred to as "Chia," an Indian term meaning "high place." Pine and scrub oak were and today still are supported on the relatively high and dry islands surrounded by swampy wetlands. This variety of habitats, host to an abundance and diversity of both plants and animals, proved to be quite acceptable to early humans.

Ortona is a site comprised of a large and significant complex of earthworks consisting of at least seven or eight large and small multipurpose mounds, at least one of which was for burials. There are horseshoe-shaped ridges, semi-circular and linear ridges, and a number of connecting causeways. Two hand-dug canals totaling about six miles in length by about ten feet in width by up to four feet deep connect the site to the Caloosahatchee and back to Fisheating Creek, thereby enclosing a large triangular area.

Human habitation of the Ortona Site may go back to the Late Archaic, with various features being utilized, developed, and added all the way up to the beginning of the A.D. 1700s. The circular-linear earthworks are of a distinctive type in the Lake Okeechobee region, occurring at Fort Center and a dozen other sites around the lake. The archaeologists concluded that:

> The site is at a strategic location in the Caloosahatchee River corridor that controls access between the river and Fisheating Creek via an interior canal route. It is possible that Ortona interacted with other Southwest Florida coast sites such as Mound Key and Pine Island as well as many of the sites around the lake. During the pre–European contact period [ca. A.D. 800–1500] the cultural and political affiliations between the Calusa and the Mayaimi of the area probably were strong.

In test diggings throughout the site, Carr et al. recovered a large number of shards of Belle Glade Plain and unclassified sand-tempered pottery, a drilled shark tooth, and a small piece of highly patinated trade copper. Several ceramic smoking pipe fragments proved to be similar to others found around the Lake Okeechobee region. Ash and charcoal associated with faunal bone indicated a number of habitation areas. Seed fragments were found of both *Curcubita* (squash) and *Lagenaria* (bottle gourd). However, nothing was recovered that might hint at practice of agriculture.

Dating back to A.D. 200, this site is significant due to its strategic location that would have controlled access between the Caloosahatchee River headwaters at Lake Okeechobee and possibly, via canoe routes, between the river and Fisheating Creek.

Today much of the site has been altered, if not partially obliterated by modern drainage, land use, sand-mining, road-building, a modern Ortona cemetery, and vandalism. Nonetheless, a number of protected features do remain and are described in situ through interpretive displays. The site is open to the public as Ortona Indian Mound Park.

Caloosahatchee Cultural Area

Pine Island

Offshore of the Gulf Coast city of Fort Meyers lies 12 miles long by two and one-half mile wide Pine Island. On its eastern side separating the

The Formative or Ceramic Period

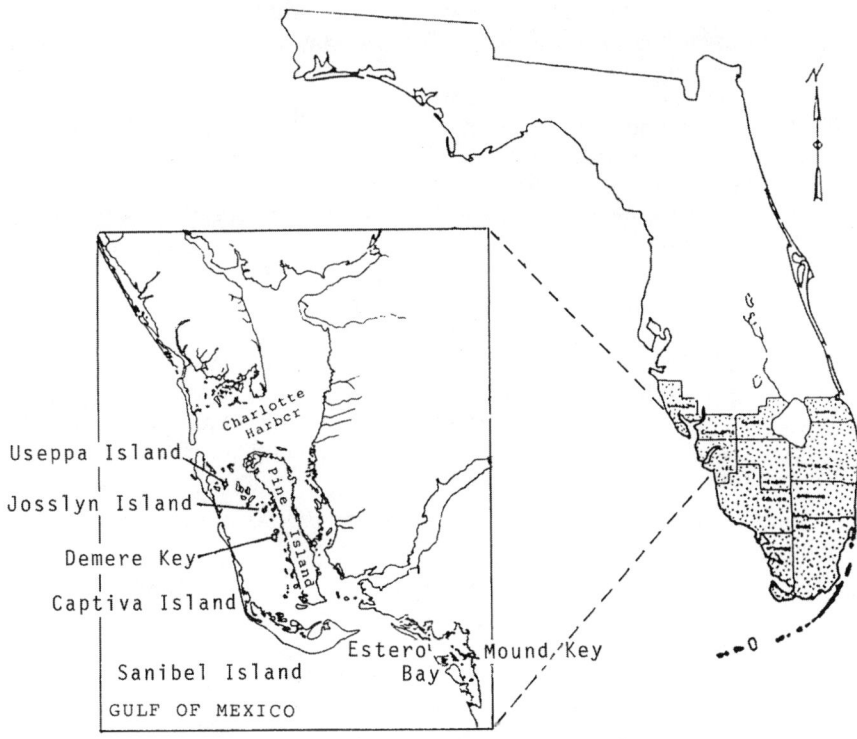

The Caloosahatchee Cultural Area centered around Charlotte Harbor and the Caloosahatchee River. Islands tucked between the Sanibel–Captiva chain and around Pine Island and the mainland afforded safe haven and abundant resources for human habitation. (Not to scale.)

island from the mainland is Matlacha Pass. Off the western shore is Pine Island Sound, within which are many small islands, some of which such as Useppa, Demere Key, and Josslyn are known archaeological sites. Between the sound and the Gulf of Mexico is a thread of protective, elongate, narrow barrier islands named Sanibel, Captiva, North Captiva, and Cayo Costa. The Myakka, Peace, and Caloosahatchee Rivers drain into Charlotte Harbor, an estuarine system that was forming over 4,000 years ago and had pretty much established its present configuration by about 2,700 years ago.

In the northern end of Pine Island Sound is Pine Island. Near the northwestern end of the island is the Pineland site, known as Battey's Landing back in 1897 when it was visited by Cushing. He notes:

> The foundations, mounds, courts, graded ways, and canals here were greater, and some of them even more regular than any I had yet seen.... The same sorts of channel-ways as occurred on the outer Keys led up to some sorts of terraces and great foundations, with their coronets of gigantic mounds. The inner or central courts were enormous. Nearly level with the swamps on the one hand, and with the sand flats on the other, these muck-beds were sufficiently extensive to serve ... as rich and ample gardens; and they were framed in, so to say, by quadrangles formed by great shell structures which, foundation terraces, summit-mounds and all, towered above them to a height of more than sixty feet.
>
> There were no fewer than nine of these greater foundations, and within or among them no fewer than five large, more or less rectangular courts; and beyond all, to the southward, was a series of lesser benches, courts and enclosures.... This settlement had an average width of a quarter mile; ... Its high-built portions alone, including of course the five water courts, covered an area of not less than seventy-five or eighty acres.

Dr. Lee Ann Newsom provides a modern description of this important site complex.

> Prominently situated and overlooking Pine Island Sound are two large mound groups known individually as the "Brown's" and "Randell" complexes.... Approximately 50 to 100 meters inland of the two large mound groups is a raised, linear topographical feature ["Citrus Ridge"] that is composed entirely of sand and intermittent midden lenses. A smaller shell midden or mounded midden ["Old Mound"] lies at the southern periphery of the site complex, and to the north and east are a large sand burial mound ("Smith Mound") and another mounded midden ["Low Mound"]. Finally, "Adams Mound" is a large sand mound, possibly another burial mound, located still farther north and east.

Human occupation of Pineland closely parallels that of many of the adjacent islands, with first inhabitants utilizing the site at least as early as the first century A.D. and up to A.D. 300. Major occupation occurred from A.D. 500 to 1600. A vertical cross section of the 29 and a half feet high Brown's Complex reveals "immense, stratified shell midden deposits interspersed with living floors, ashy activity areas and hearth-like features ... closely linked to the Randell Complex which seems to have shared structural and developmental history, including episodes of intentional mounding and redeposition of midden materials," writes Dr. Newsom.

Geologists collaborating with a University of Florida archaeological research team determined that at least two Pineland middens, neither of which is described above, originally deposited on dry land between the years A.D. 500 to 900, now are inundated by seawater at the lowest level of the site. Obviously, sea level was lower during that period of habitation.

Transecting the island and the Pineland site complex to Matlacha Pass is a canal believed to have been hand-dug by a large labor force of the Calusas. During his 1895 travels as he worked his way through the domain of the Calusas, Frank Cushing wrote that the canal still was 30 feet wide and eight feet deep. The years during which this engineering marvel for its time was undertaken are unknown, but likely started back during earliest site occupation when the Gulf of Mexico's level was a bit over three feet higher than today. As previously noted, from A.D. 500 to 900 the water level was lower and work would have been more difficult, necessitating a deeper canal. The water rose again from A.D. 900 to 1400, and receded thereafter during the time known as the "Little Ice Age." Either or both high water level periods would have made canal-digging easier, lower levels requiring a much deeper excavation. The low water periods may explain the unusual depth of the canal. Executing a project of such broad scope is considered a mark of the Calusas' cultural complexity. Archaeologist George Luer has suggested the canal was dug by the Calusas after A.D. 1,000, about midway through the site's major occupation, to transport materials from the mainland. In any case, it obviously was a massive undertaking by a large number of people. The fluctuations in sea level are corroborated by evidence from nearby barrier islands.

Though a major goal of archaeology is to explain the past, it should also help to explain contemporary events and possibly enable prediction of future events. Thus, reporting the site in *Calusa News*, archaeologists Walker and Marquardt state, "If such fluctuations can be identified in prehistory and observed as cyclical, then they also may be predictable."

The site's wet earth maintained anaerobic conditions within the middens that preserved wooden debris, a finely carved wooden bird head, twisted palm fiber cordage, and seeds of chili pepper, papaya, and several types of wild gourds and squashes. Deer bone artifacts included compound fishing hook elements in a remarkable stage of preservation. Post molds were confirmed in four locations, including a linear arrangement of postholes that documented the first prehistoric structural floors ever found in southwest Florida.

Regarding recovered pottery, the authors write:

> Pineland is justifying its reputation as an extraordinary archaeological site. This is nowhere more evident than in the remarkable diversity of decorated pottery coming to light as the dig progresses. When one considers how many pottery making traditions are represented here, and

from how far away they must have come, Pineland emerges as a center of major significance in the aboriginal world of South Florida.

Among the many pottery types found are Matecumbe Incised, Weeden Island Punctated, Little Manatee Zoned Stamped, Belle Glades Plain, Ruskin Dentate Stamped, Pinellas Plain, and St. John's Check Stamped. The Glades Plain was made from after 500 B.C. up until the 1500s A.D. It is undecorated, of sandy texture and often referred to simply as sand-tempered plainware. It changed little during its 2,000 year history and, so is difficult to use in determining site age. Ceramics technology expert Ann Cordell, Senior Biological Scientist of the Florida State Museum, has utilized microscopic analysis of thousands of plain-ware pottery shards from southwest Florida sites (Useppa Island, Cash Mound, Josslyn Island, and Buck Key) to discover the kinds of clays used by the Indians. Sorting the shards into categories of kind, size, and relative constituents in the clay, she has found far greater variability than meets the unaided eye. Categories based on relative abundance of sponge spicules, quartz sand, and other mineralogical properties in the clay from which the pottery was made sort it into three varieties corresponding to time periods of Indian culture in southwest Florida. A collection of clays from different areas, formed into briquettes and fired at varying temperatures, provide comparative samples to determine the ancient potters' material sources. This helps determine whether the pottery was made locally or had been imported.

Seasonality studies help determine the time of year a site was occupied. Many animals display a clear seasonal growth pattern in their bones or shells. Study of the animals remains can determine at what season they were collected, thereby indicating when people lived at the site. Deer, oysters, quahog clams, scallops, catfish, pinfish, and thread herring are reliable examples. If it has been determined that in a particular area scallops had been gathered in summer, quahogs in the winter or spring, and catfish, pinfish and herring in the fall, one might deduce that the site was a permanent, sedentary community (as was found to be the case with Horr's Island).

Here the archaeologists who are concerned with how the people adjusted to their changing environment share mutual interests with geologists. The latter study rock, shell, plants, pollen grains, sand, and clay in core samples taken near archaeological sites. These provide a record of temperature, rainfall, storm frequency, and sea level fluctuations. The "geoarchaeologists" are able to determine sea level fluctuations by the position of ancient habitation sites in relation to current sea level. Thus armed, the former can understand environmental conditions and the effect it had on the people and their culture throughout time. Of course, the foregoing is but one example of interdisciplinary effort. Place the prefix "archaeo" before just about any scientific discipline such as zoologist, biologist, chemist, or ecologist and you have the idea.

Pine Island's excavations also produced a large enough single site sample of 1,400 shell artifacts from different time periods to permit them to be organized into time units, thus revealing their change, if any, over the 1,500 years of occupation. Among them are fishing sinkers, a rare horse conch columella woodworking plane, columella gouges, a hafted fighting conch hammer, shell net-mesh gauges, surf clam and Venus clam knives, and fish scaling tools. The archaeologists believe that shell artifacts displaying changes in shell tools may relate to increasing complexity among the Calusas.

Formerly hidden beneath ground level due to topographic changes over the years, Pineland's mounds now are of continuing scientific and public interest. Many thousands of volunteer hours have been expended during the Florida Museum of Natural History's site field seasons, hundreds of teachers have received special training, and thousands of school children have enjoyed on site contact with archaeology. The work continues.

Owners of the Pineland Site, Don and Pat Randell, generously donated about 56 acres of their property to the University of Florida Foundation, where the museum is establishing the "Randell Archaeological Research Center" for continuing research and educational programs at Pineland. In recognition of their contributions, the Florida Archaeological Council has presented the Randells with a Steward of Historic Preservation award.

A member of one of the early Spanish expeditions named Garcelaso left a written record of a town referred to as Ossachille. While it may not be a description of Pineland, it is a rare insight into the appearance of a sizeable Calusa village at the time of early European incursion:

> As many as possible, at least the Chief and under lords, had houses built on the flat top of the mound site, according to the grandeur of the ruler, accommodating from ten to twenty houses, for the dwellings of the family and the serving people. On the flat at the foot of the hill [mound] they make a quadrangle square, according to the size of the village which is to be located around it.

The town of Ossachille, thought to have been located in today's Manatee or Sarasota County, might be considered to represent larger settlements of people in any of the sites throughout the Calusa domain.

Useppa Island

Within the Caloosahatchee Cultural Area west of the Pineland site in Pine Island Sound, islands are typified by shell middens and shell works, some of which are of gigantic proportions encompassing entire islands such as Useppa, Josslyn, Horr's, Demorey, and Mound Key.

Some twenty miles west of Fort Meyers is Useppa Island. It holds the distinction of having achieved the highest elevation in its county, due to the aboriginals having built upon a naturally high mainland Pleistocene sand dune and laying down their shell, ash, and refuse almost continuously for 5,000 years to the historic period.

As sea level rose and some mainland features became islands, earliest habitation sites, like those on Pine Island, probably were submerged. People had to move from place to place not only due to the slowly altering topography, but as necessitated by their lifestyle of fishing, hunting, and gathering as well.

Here, as with parallel work on other southwest Florida sites, studies of recovered quahog clam shells, odostome shells (tiny snails that adhere to the outside of oyster shells), catfish otoliths, and atlas bones of pinfishes, grunts, and thread herrings enabled determination of season of the particular animal's death, thereby indicating seasonality of site occupation.

Stratigraphic and radiocarbon-dating studies of the high Useppa Island Mound revealed Indians were living on the island as early as about 3875 B.C., 500 years before Egypt would be ruled by its first pharaoh. The site served as a collection and processing location for the early shell tool industry. From about 3,000 years ago until the early 19th century, it was utilized by a succession of people whose vague history is outlined in the midden layers context. In addition to pottery shards, some strata were characterized by large amounts of fish bone, others were separated by layers of crushed sea urchin spines and shells. Obviously, a single species from time to time was collected in great numbers. On other area mounds, dense deposits of a given species show that certain localities produced them in abundance and they were exploited by the people. There also were the "usual" conch and whelk celts, hammers, cutting tools, awls, and bone points. Additionally, sinkers and net-mesh gauges were found, and one of the bone points clearly seemed to be part of a composite fishhook. Obviously, these latter three items indicate knowledge of cord-making by the twisting of vegetable fibers. Associated with several definite work stations was the debitage, the debris, resulting from manufacture of such tools, proof of early man's ability to mass produce quantities of necessities in advance of actual need. What proved to be unusual about these artifacts was that they were shown to indicate a complete linear sequence of tool–making on what was a workshop floor dating back 3,350 years ago. Apparently, shell tool manufacture was established in the Late Archaic period. The people were learning to master the sea and its bounty.

A single well-preserved articulated skeleton was found buried in the Useppa Island midden. This male had been in his 30s and stood about five feet seven inches tall. Radiocarbon dating revealed that he had died around 1,400 years ago; rather than being of the much earlier Archaic Period context, he simply had been interred deeply within the mound at that later time. Since his are the only remains found in this location, it cannot be presumed that his people had achieved his tall physical stature. However, I. Mac Perry (1993) does describe physical characteristics of aboriginals of that time as "males averaging about five feet six inches in height, females five feet; muscular and a little heavy; their teeth were severely worn, often to the gum, from gnawing bones, eating gritty shellfish, and using their dentition as a fifth appendage."

Forensic dentist Dr. Richard R. Souviron comments regarding ancient human teeth upon which he had worked: "They had been worn right down to the gum due to the dirt in the food they ate."

"Dr. Souviron is quite correct in his assessment of the extreme dental wear seen in the teeth of prehistoric human populations in Florida. Indeed, I have commonly seen molar teeth so worn that the crown was worn away leaving the roots (two in the case of lower molars and three in the case of upper molars) remaining as small separate teeth. In many cases, apical abscesses developed after such extreme wear. On the other hand, complete loss of teeth is seldom seen in these skeletal samples," wrote the late Dr. William R. Maples, curator-in-charge, the C.A. Pound Human Identification Laboratory, Florida Museum of Natural History, Gainesville.

How those people were able to withstand what must have been extreme pain resulting from such excessive wear is explained thus: "As the crown wore down, the nerve receded beneath its surface until eventually the tooth or teeth died; thereafter the wear of the dead tooth could continue right down to, even beneath the gum." Robin C. Brown (1994) writes that as well as their dental problems, they and their forebears suffered arthritis, especially in their backs, from syphilis, rickets, wound infections, and birth defects.

Just about the time of the burial of the man in Useppa Mound, use of the bow and arrow had reached the North American eastern woodlands. According to Dr. Barbara Purdy (1996), "The bow and arrow was first in use 19,000 years ago in North Africa. It is believed that this technologically superior weapon was invented only once and that it eventually diffused throughout most of the world." Assuming her conclusion to be correct, it took nearly 18,000 years before the concept reached all the way to the southeastern United States. This efficient weapon was added to the Florida aboriginal arsenal, but did not entirely displace use of the atlatl throughout the peninsula. One ship captain who arrived in the Keys in the 16th century noted in his log that the natives were very accomplished both as archers and with the dart, the latter probably a reference to the atlatl's short spear and confirmation of its continued use. Lack of appropriate stone for arrowheads or dart points posed little difficulty. Deer antler points, bone, wood, stingray barbs, horseshoe crab tails, fin bones, fish teeth, fish scales, shell, teeth, claws, turkey cockspurs, viper's teeth, bird bills, anything would do. Cane shafts cut to a fire-hardened point were capable of piercing just about anything; later, even chain mail Spanish armor.

Josslyn Island

Moving south from Useppa Island, Cushing's report on Josslyn Key, as one might expect, is rather expansive. He describes

> a remarkable central court of less than half an acre in extent. Five very high and steep mounds ... led forth from the court diverging to the sea [and] formed its western and southern side, while its opposite side and end were formed by two extensive platforms, also exceedingly steep within and nearly as high as the elevation, and dividing from these and from each other by straight canals that led forth in northerly directions far out through the mangrove-covered enclosures down toward which the platforms were terraced. The court is very deep and so regular that it resembles the cellars of an enormous elongated square house.

The canals had been dug to afford access by canoe from the village to Pine Island Sound. He also notes the steep face of the mound is reported to have been "faced with conch shells driven into the shell bank, leaving the large ends exposed."

Horr's Island

This island, which was discussed earlier in chapter 3 (pages 56–58), was visited by John Kunklel Small, who reported that while there are high sand dunes on the western part of the island, on the eastern part facing north there is said to be a burial ground and a series of still higher kitchen middens and shell mounds which fall off abruptly into Caxambas Bay. The perpendicular mound walls, which clearly display strata of different shell species, are the result of shell removal for road building. There also are layers of charcoal and animal bones, all of which indicate what was eaten. He writes that "prodigious quantities of conch shells, each one punctured in the same way for the purpose of removing the animal for eating."

Small notes that various stone implements, obviously imported from the north, have been found in the middens, including a foot and a half diameter circular millstone with an eccentric hole, evidently for a handle. He also writes that two "beautifully finished ceremonial ax-heads of granite or syenite have been unearthed, which, of course, had to have been imported."

Mike Russo's team, working with the Southwest Florida Project, determined that a conical sand and shell mound was separated by about 3,500 years from later typical Calusa mounds constructed of debris, and layers of dirt, ashes, bones, and shell that served as house foundations. Mound A had been carefully constructed on top of a sand mound in deliberate layers of sand spread over shell. Some of the sand layers had been colored by addition of charcoal. Two human burials on its fringe lead Russo to believe this mound served ceremonial use.

Mound Key

South of Pine Island Sound, the afore-mentioned Keys, and the southern tip of Sanibel Island is Estero Bay. Within the bay, off the southern end of Estero Island (present day Fort Meyers Beach), is 125 acre Mound Key. Most of the island is designated by the Florida Park Service as Mound Key Archaeological State Park and is managed by Koreshan State Historic Park. The island certainly lives up to its name. The Park Service history of the site states:

> Sustenance of the Calusa was composed primarily of shellfish and fish, which is apparent through composition of the mounds that are the structure of the island.... Mounds were constructed by the collection and organization of "midden" which is a collaboration of shells, fish and animal bone, and artifacts such as pottery ... their composition is made from waste products of their culture ... intricate compositions of substrate that were used for display of power, religious monuments, and as burial memorials.

Mound Key is thought by many to be the site of Calos, the main community of the Calusas, and the seat of the casique Carlos. It is a largely artificial island dominated by a 30 foot high cone-shaped platform mound that today is topped by a 75 foot diameter flat area. The balance of the island is covered with a multitude of smaller mounds ranging in height from five to 15 feet. A number of those are felt to be dwelling mounds and at least one is a burial mound. Frank Cushing writes:

> One [key dweller's village] was at Mound Key or Johnson's Key as it was variously called. I make mention of my visit to the place principally because of its great extent. It consisted of a long series of enormous elevations crowned by imposing mounds and reaching an altitude of over sixty feet. They were interspersed with deep inner courts, and widely surrounded by enclosures that were threaded by broad, far reaching canals, so that this one key included an area of quite two acres.... I was told by Mrs. Johnson, who was the wife of the owner of the place ... that burial mounds ... occurred in the depths of the wide mangrove swamps, that lay between the mainland....

The island is almost divided by a canal that is about 50 feet wide and narrows to about four feet in two places. It fills with seawater at high tide and the narrows are thought to be places at which fish were netted. Or, weirs of wood or brush might have been placed at the narrows to contain seawater and the fish that came in with the tide. The living fish could be held in the containment until needed. Lowering the water at low tide left the fish stranded, to be gathered with minimal expenditure of energy. The bulk of the key, in fact, floods with the incoming tide, creating a labyrinth of pools and canals, among and above which stand the smaller dry mounds.

Today, mangroves and vegetation restrict the view of the complexity of the key, but an aerial view shows that smaller canals spread in every direction from the main one. Mound Key is considered by some to be the most elaborate occupation site of the early people and their antecedents. They were concentrated on the Gulf Coast from Charlotte Harbor south through Marco Island from about 500 B.C. to A.D. 1500.

The people of the Gulf Coast laboriously dug their canal systems with shells and baskets for a number of very sound reasons. One canal that is 30 feet wide and four to six feet deep, traverses the two and a half mile width of Pine Island. There is evidence that it continued seven and a half miles across Cape Coral to Yellow Fever Creek, which joins the Caloosahatchee River, in turn connecting with Lake Okeechobee. Another canal had been dug at the headwaters of the Caloosahatchee to circumnavigate the riff that existed where the lake spilled into the river.

Canals were dug to afford access by canoe from the village to Pine Island Sound. Just south of St. Petersburg, Ruskin Thomas Mound is near the mouth of the Little Manatee River. Its canal is about 64 feet wide at its beginning at the southwest side of the mound, and runs for 238 feet, tapering down to 30 feet in width where it joins the water. In Koreshan State Park east of Fort Meyers, there are two huge mounds, one 28 feet high and the other 40 feet high, that once were connected by a canal discernable today only as a rather swampy area. About 100 years ago, Kenworthy described the Gordon's Pass Canal in present day Naples in the Smithsonian's Annual Report. The Indian Canal site on the south side of Naples once connected the Back Bay area to the Gulf. Caxambas Mound is a complex of shell middens covering some 50 acres on the south end of Marco Island. Canals had been constructed between the mounds and leading to the Gulf. Goodland Mound, on the south end of Marco Island, had a canal connecting two mounds that actually continued in use through the early 1900's. Addison's Place is a site consisting of 30 acres of shell formations located about five miles east of Key Marco. The shell forms north-south parallel ridges. Between them is a canal connecting the water outside the Key to a triangular pond. Another permits water access from the Gulf to two large shell mounds at the northern approach to Chokoloskee Island. The canals served to simplify canoe travel from mound to mound, from mounds to outside waters, even to separate the living from the dead. They also indicate that life had become more sedentary, and that food gathering no longer dominated almost every waking minute of the day.

High winter winds did prove hazardous to canoeists in offshore waters. However, the naturally protective island barriers provided sanctuary within the bays, and as Chuck Blanchard writes in a 1995 issue of *Calusa News*, "In no other area in South Florida is such dependable water travel over such a large region possible. It is no wonder that sites abound within its borders."

Man improved upon nature's waterways by digging his canals, and area cultural traditions became more and more complex for no less than five millennia. The religious and political organization of the bands or chiefdoms had reached a state of advancement that permitted planning and execution of such major projects.

As noted before regarding Tampa Bay, this entire Gulf Coast region was forming 4,000 years ago. Archaeologists have provided data showing that sea levels, vegetation patterns, and estuaries had reached current levels by about 700 B.C. This rather unique coastal/estuarine environment evolved into a virtual Eden that appears to have provided just about every prehistoric human necessity for thousands of years.

On Mound Key pottery shards are in abundance, and a number of other typical shell artifacts have been found. Years ago, two huge shell middens on the island were excavated for road construction. Contractors dug out shell material to six feet below low tide line. They said there was a lot more shell below that, leading them to speculate that the key had been occupied for many thousands of years. Earliest inhabitants of the site must have started disposing of their shells when sea level was six or more feet lower than today.

In Widmer's 1984 Ph.D. dissertation, he summarizes that these people chose a sedentary existence which led to increased fertility and rapid population growth. By 1,200 years ago, a high level of population necessitated a centralized power structure in order to resolve disputes and redistribute food and other resources effectively. Hence, the evolution of the chiefdom. Renfrew and Bahn (1991) define a chiefdom as a society operating on the principle of ranking; i.e., differential social status. Different lineages are graded on a scale of prestige, calculated by how closely related one is to the Chief. The society generally has a permanent ritual or ceremonial center, as well as being characterized by local specialization in crafts. Surpluses of these and of foods are paid as obligations to the Chief.

The lifestyle at this site remained essentially unchanged to the time of contact with the Spaniard Pedro Menendez in about A.D. 1556.

A human profile drawn from comparison of remains is "one of generally consistent physical characteristics and health for prehistoric populations throughout Florida from the late Archaic Period through European contact," according to archaeosteologist Amy Felmley of the Pine Island site. This certainly is consistent with a current theory of human origins stating that human evolution seems to be characterized by long static periods during which there is little change.

Archaeologist Jim Lord holds a very large horse conch shell in his left hand. Based on its size and weight, and the holes in the outer shell whorl, he feels it most likely served as an anchor. His right hand holds the columella from an equally huge conch. This piece of solid, heavy shell served well as a tool. Both artifacts were found in the Marco Island area.

From approximately 700 B.C.[6] until well after European contact in the late 15th and early 16th centuries, the people whose history is revealed in the deposits probably were Calusa or their antecedents. Many Spanish relics found on site attest to early European residence. Again, Cushing writes:

> Spanish relics have been found such as Venetian beads, scraps of sheet copper, small ornaments of gold and silver, and a copper gold locket.... It contained a faded portrait and still more faded letter, written on yellow parchment apparently from some Spanish Grandee of about two hundred years ago [1696] to a resident colonist at that time.
>
> Whether these relics indicate that here the ancient key dwellers or their intermixed descendants had lingered on into early historic times and that the Mission that these betokened had been established among themselves or among alien successors, could not of course be determined.

Many anthropologists today concur that Mound Key in Estero Bay was the site of the capitol of the Calusas. They base this on the translation of an account written by Lopez de Velasco, who described the island as being "two miles in circumference lying in a bay fourteen to eighteen miles in circumference." In 1566 Menendez had a subordinate ascertain from the Indians that a river flowed from the Miami Lagoon (Lake Okeechobee) two leagues to the town site. The river in question would have been the Caloosahatchee and the distance from lake to Estero Bay or Pine Island Sound would be about right. That is a far better match than descriptions of Charlotte Harbor and any of its islands a bit farther north.

Well armed with atlatl, spear, harpoon, bow and arrow, and throwing sticks it would seem just about anything was fair game. But studies indicate that these people, the Calusas, relied more on an aquatic subsistence. From the incredible size and volume of their shell mounds, it is easy to suggest that shellfish was the major dietary contributor. However, animal remains analyses show that fish were the major contributor. Shellfish, turtles, other reptiles, mammals, and birds accounted for a very small percentage of meat weight.

Site archaeobotanical analyses have failed to yield any domestic plant species that might indicate that some form of agriculture augmented the hunter–gatherers' diet. Even the earliest Spanish records note flatly that the Calusas and other people of the south part of the peninsula practiced

---

[6]*The Mayan civilization in southern Mexico, pre–Inca culture in Peru, and Mississippi Valley culture in North America were forming. Basilicas were being built in Rome and first plans for the Vatican Palace were drawn.*

no form of agriculture. Apparently there was little need to labor in fields since the environment provided an abundance of wild edibles, including botanicals, there simply for the taking.

TEN THOUSAND ISLANDS CULTURAL AREA

Big Cypress

The Ten Thousand Islands Cultural Area encompasses today's Big Cypress National Preserve, within which almost 400 archaeological sites have been recorded. They vary from temporary campsites to large settlements of 100 acres or more. Sixty-four radiocarbon dates have determined habitation ranging from 3,200 years ago. Within the Big Cypress, the people took advantage of the freshwaters that gently course their way from Lake Okeechobee to the Gulf of Mexico, and the bounty of wildlife produced therein and in the surrounding environs. Farther west the multitude of islands of the Gulf Coast were most intensely occupied where the people easily could take advantage of the richness of the sea, as well as of the estuarine environments created by drainage from the great lake.

Key Marco

Earlier efforts of Frank Hamilton Cushing inspired him to write that the islands throughout the area "evidenced remote aboriginal occupation … their condition and their occurrence beneath the peaty deposits of muck might even betoken some such phase of life in Southern Florida as that of the Ancient Lake Dwellers of Switzerland, or of the Pile and Platform Builders of the Gulf of Maracaibo, or the Bayous or the Orinoco in Venezuela." Hence, this Key Marco site is known as the "Court of the Pile Dwellers."

This initial commentary was affirmed when his excavations at Key Marco produced an amazing array of artifacts, the likes of which have yet to be uncovered elsewhere on the southern half of the Floridan Peninsula. There were timbers and decayed thatch, plaster of hearths and charcoal from domiciles. Pottery bowls, cooking pots, mortars and pestles, net weights and sinkers, fishhooks of pointed bone, shark tooth knives, their handles carved to fanciful animals, clubs of wood embellished with split wolf jaws, and atlatls. Wooden objects still bore red, gray-blue, black and white paint. Domestic artifacts included wooden boxes, trays, bowls, mortars and pestles. Artistically carved and painted wooden masks, a unique horned deer's head and wolf heads, a kneeling panther figure, a ceremonial baton, and a toy dugout catamaran were among very special riches. Cushing's respect for the ability of the ancient artists is evident in his statement about the carved and painted deer's head:

> This represents the finest and most perfectly preserved example of combined carving and painting that we found.... In form, or mere contour, it portrayed with startling fidelity and delicacy, the head of a young deer or doe.... [The ears] were also relatively large, and were fluted, and their tips were curved as in nature, only more regularly; they were painted inside with a creamy pink-white pigment to represent their translucency; and the black hair tufts at the back were neatly represented by short black strokes of paint, laid on lengthwise and close together.
>
> The tortoise shell eyes still remained in place, and the combined beeswax and rubber gum cement with which they had been secured was still intact when the specimen was found. The whites of the eyes had consisted of some very bright gum-like substance, and the front corners or creases of the eyes had been filled with black gum and varnish, highly polished, so that, save for the conventional sets of equidistantly radiating winter-marks they gave surprisingly life-like realistic and timid or appealing, yet winsome, expression to the whole face. The muzzle, nostrils and especially the exquisitely modeled and painted lower jaw, were so delicately idealized that it was evident the primitive artist loved, with both ardor and reverence, the animal he was portraying.

He writes on, describing the artisan's method of ear attachment in such a manner as to allow their movement and the method of fixing deer hide to the back of the mask

> the more perfectly to disguise the actor who no doubt endeavored in this disguise to impersonate the character of the deer god or dawn god, the primal incarnation of which this figure was evidently designed to represent.

Implements included shell axes, gouges, and hammers, Horse Conch drinking cups, and handsomely crafted adzes. Of the latter, its twelve-inch long, hard wood handle angled sharply as a carved form of animal head, the mouth of which was a deer horn receptacle within which the adze blade was affixed. Shark tooth knives, some of which still bore the wooden handle, were plentiful. Shark teeth also lined the length of a hard wood saber club. Knobbed bludgeons were found, as were a number of both one and two-holed atlatls, the finger holes designed to improve the thrower's handle grip.

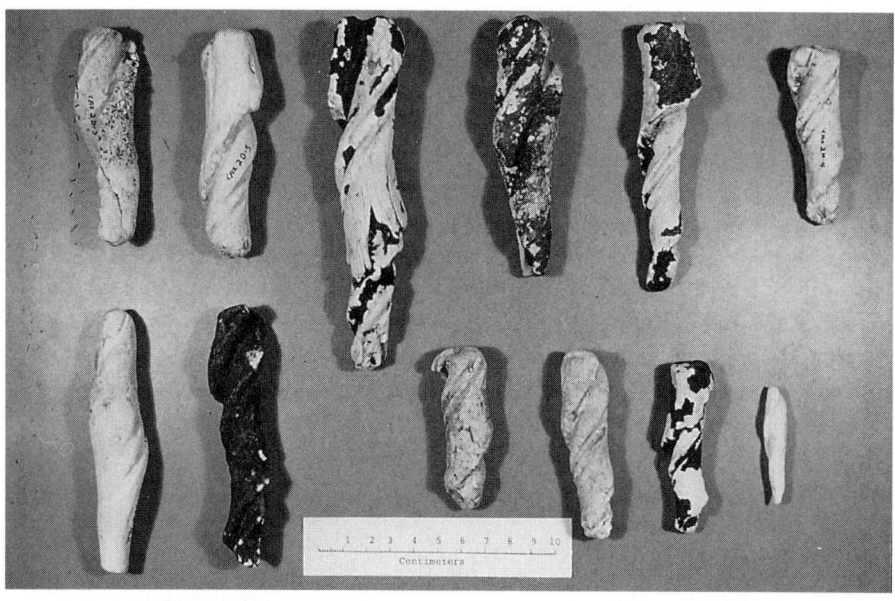

Marco Island artifacts. Columellae from various species of shell served any number of utilitarian purposes — hammer, punch, awl, gouge or whatever.

A collection of shell celts shows various stages of wear on the sharpened wide end.

Clusters of shells strung on twisted fiber line served as anchors or weights for fishing nets. Olive shells with perforated ends might have been used as necklace beads. The shark tooth at bottom right has a broken point and worn cutting edges, indicators of use as a knife and drill.

Gourd fragments and fiber cordage had been preserved in the anaerobic muck between the site's higher mounds and shell ridges. Most of the organic fiber materials (the study of which today is referred to as 'soft archaeology') including rope, netting, and cord that fastened float pegs to the net, disintegrated soon after contact with air, but not before being sketched by an on-site artist. Much of the wood, stone, shell tools, and pottery were safely curated and are on display in the Smithsonian, the

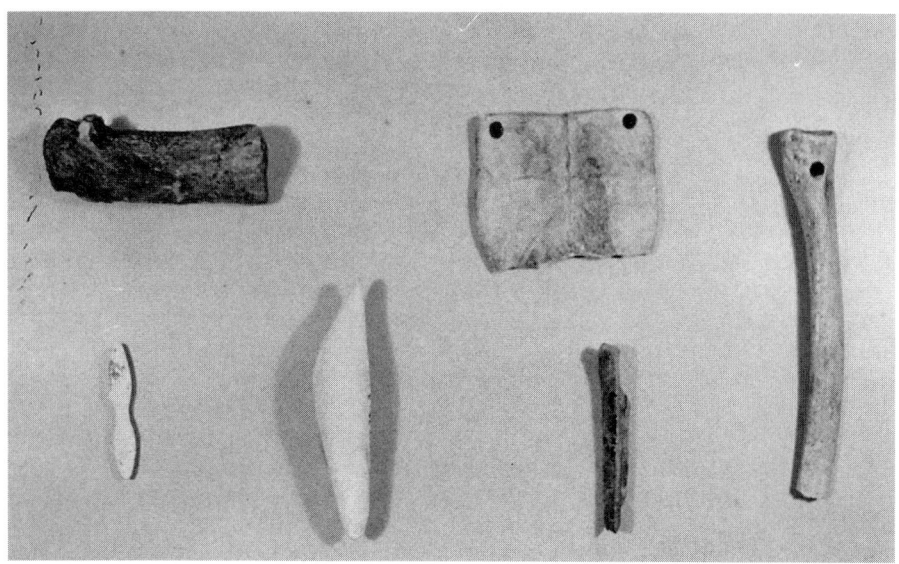

People of Marco found bone useful as well. The square, six sectioned, drilled piece at top center probably adorned someone's chest as a gorget made of turtle plastron.

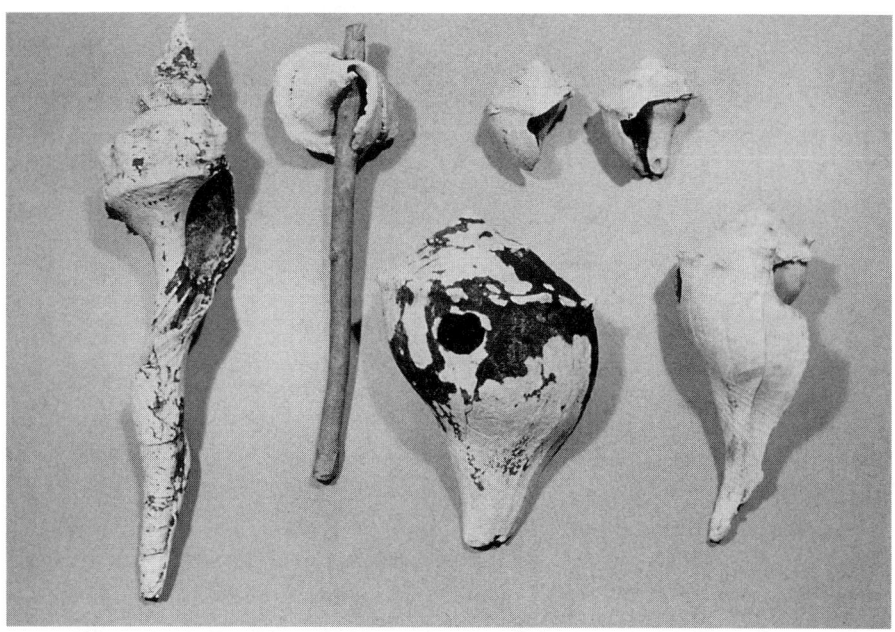

Shells were hafted to be used as an axe, adze, or hoe. With the outer whorls broken away, it became a hammer or club.

The Formative or Ceramic Period

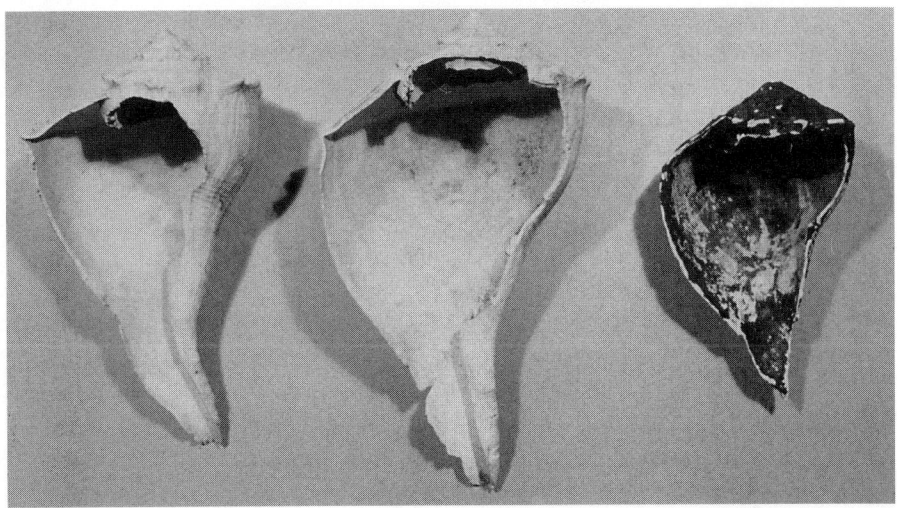

Shell dippers or cups were made by removing all but the outer whorl. Some are found with burned areas on the outside, proof that they had been placed on a fire to heat or cook whatever they contained inside.

Shell pendants adorned aboriginals of Marco Island. They were suspended around the neck, arms, or legs on thongs or twisted fiber chord. (Marco Island photographs: Florida Department of State, Division of Historical Resources, Bureau of Archaeological Research, Tallahassee.)

National Museum of Natural History, University of Pennsylvania, and the Florida Museum of Natural History, University of Florida, and other museums today.

Through the ensuing years other researchers have followed. Their work at Key Marco has reinforced Cushing's and established the area's archaeological uniqueness. This has led archaeologists to believe that Key

Marco apparently played a very special and important role in the religious activities of the people of the lower Gulf Coast.

Key Marco may have been the source of the most impressive archaeological finds in the Ten Thousand Islands area. However, other mounds exist at the mouth of Lostman's River, Lostman's Key, Johnson and Hamilton mounds, Gopher Key and the site of the Watson place on Chatham River. On the east bank of Turner River, just a short distance upstream for roughly 1,000 feet, is the Turner site, a complex of 30 or so row mounds ranging in height from 12 to 20 feet. One hundred-fifty acre Chokoloskee Island, all shell and sand, was described by a visitor in 1900 as bearing a shell mound 27 feet high above low tide at the island's northern approach. A canal running in from the southern shore terminated in two prominent mounds. In 1972, a contractor, after dredging and grading at the north end of the island, related that the shell extended about ten feet below mean water to bedrock. It was also noted that "gumbo limbo trees mark ancient habitations, seemingly posting themselves without fail as sentinels over abodes of departed people."

Bill Marquardt, curator of Archaeology and director of the University of Florida Institute of Archaeology, also serves as project director of the "Southwest Florida Project" for the University's Institute of Archaeology and Paleoenvironmental Studies. He is charged with researching the Calusa domain extending throughout the Charlotte Harbor-Pine Island Sound area of the southwest Florida Gulf Coast, including Big Mound Key, Galt Island, Cash Mound, Useppa Island, Buck Key, Wightman, Pineland Site Complex on Pine Island, Josslyn Island, Demere Key, Mound Key, and more. He states: "At the time of European contact in the sixteenth century, southwest Florida was the domain of the Calusa, a politically, socially, and culturally complex society." Political alliances, hence influence of the ruling casique, extended throughout much of southern Florida.

"The Calusa Domain represents an extremely important source of information about adaptation and cultural change in rich, subtropical settings." Subsistence of the people was based upon the naturally abundant productivity of the coastal and estuarine environments in which they thrived, quite possibly without benefit or necessity of horticulture. Hence, they are known as fishers, hunters, and gatherers. In *Man of the Everglades*, Charlton Tebeau writes that the Calusas and Tequestas

> were at home on the lower peninsula. There they developed a way of life considerably different from Indians of Northern Florida, but anthropologists generally agree that they are of the same origin. Comparative

isolation and differences of climate and natural resources are sufficient to explain the distinctive features of the South Florida Indian culture.

Both groups relied upon the natural food resources of the area, primarily shellfish and fish supplemented with game and wild plants. In few other areas could people live year round by fishing, hunting, and gathering wild plants ... the greatest concentration of this food supply was on the west coast, and there the Indian culture reached its greatest richness and diversity.

As the land provided, the people settled in the most advantageous locations and populations seem to have been regulated by the area's ability to provide through seasonal and long term climatic fluctuations.

Much gathering of detailed archaeological and environmental data remains to be done: site chronologies and settlement characteristics; human demography, environmental capacity, fluctuations, and degrees of patchiness; and climatic changes and their effects. That is a tall order even for such an in-depth study of the fascinating people of the Calusa domain.

THE EVERGLADES CULTURAL AREA

The Everglades Cultural Area encompasses millions of acres of wetlands, much of it considered uninhabitable, at least on long term basis. This dynamic environment frequently is altered by seasonal dry weather fires and fluctuations in volume of freshwater flow. Water was relatively abundant year-round along sloughs in the upper glades. Sawgrass (*Cladium jamaicensis*) grew to twice the height of a man and as it died and decayed, it added constantly to the bed of thick, rich muck that in some areas reached a depth of a dozen feet or more. Farther south in the seasonally dry lower glades the same species of grass anchored in thin soil over rough bare rock might barely reach a height of a few feet. Shark River Slough, ranging in width from 40 miles, varying in depth from three feet or more during the height of summer rains to virtually nothing except in its deepest channels in dry times, is the primary water channel for 90 miles from the great Lake Okeechobee to Florida Bay. Farther east the other major slough draining slowly to the Bay is Taylor. Land elevations from silt deposition can alter growing conditions to the point where a mangrove stand eventually will rise above surrounding wetlands and become a buttonwood hammock, or erosion may reverse the situation. Severe tropical storms often alter the coastline by closing inlets or opening new ones, reducing beaches in one area only to build up others elsewhere, denuding hammocks or introducing foreign invasive seeds, plants, even small animals.

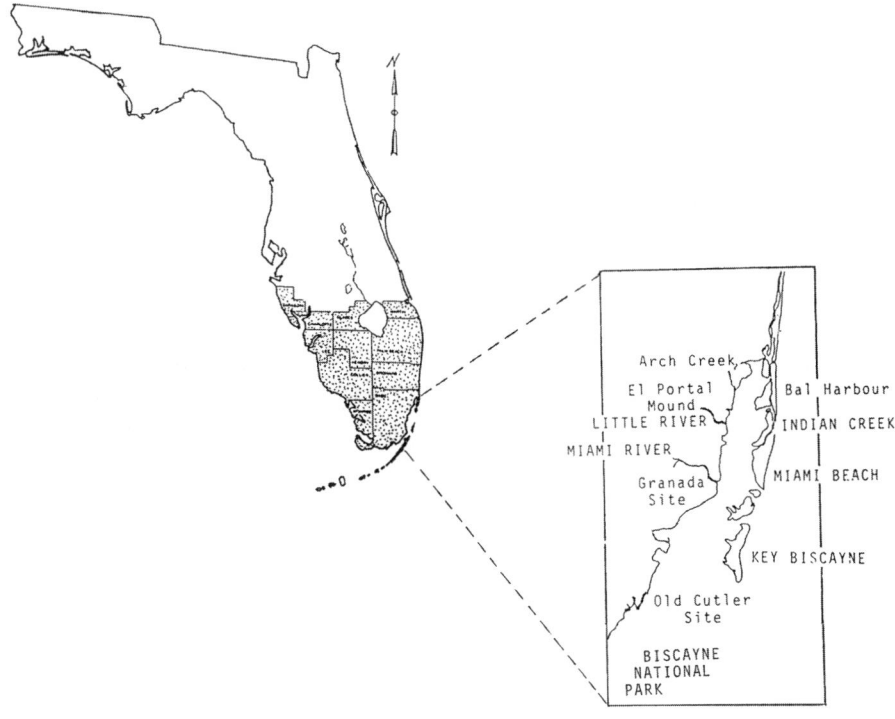

The Everglades Cultural Area supported the bulk of its human population adjacent to rivers, streams, and springs, the source of which was subterranean flow from the glades to the west. Offshore islands offered protection from heavy seas, alternative resources, and access to the richness of the ocean beyond. Only those sites referred to in the text are shown. Many more are recorded, with some, such as Arch Creek, Matheson Hammock, Everglades and Biscayne, protected as county or national public parks. (Not to scale.)

Since humans first explored the Everglades to this modern day, the environment has proven impossible to establish permanent residence. In the thousands of years before mans' canals and levees reshaped the Glades, the marshes and sloughs covered more than a third of the 6 million acres of wetlands lying south of the great lake. It is known that man survived for varying lengths of time in the Shark Valley Slough area near today's Tamiami Trail (U.S. 41), at Cape Sable, throughout the Keys, and along the Atlantic Ridge and its offshore islands.

Along the Gulf of Mexico border of Everglades National Park below Marco Island are numerous shell mounds. There are two on the Chatham River about 40 acres in extent, and 60 acres of ridges and mounds are on

Dismal Key. Others are to be found on Rookery Mound near the head of Shark River, Castaway Mound south of Chatham River, the Turner River site on the river's south bank near the Gulf, Sandfly Island, and Russell and Gopher Keys. The latter is huge and quite possibly is the least disturbed in the park. The National Park's Lostman's River ranger station was constructed on top of a mound. South of these difficult wetlands and keys is Cape Sable.

Cape Sable

Cape Sable projects into the Gulf of Mexico at the southern end of the Ten Thousand Islands Cultural Area, the very southernmost point of the continental United States. Offshore winds and savage ocean currents rendered the area somewhat less than hospitable to primitive man. The glades' ubiquitous insects had to have compounded animal as well as human misery. Herbert Job writes in 1902 about attempting to camp there for a week:

> Poor forlorn country! Though the soil is suitable for raising tropical fruits, the lack of freshwater and terrible insect scourge make it simply torture to stay there. Clouds of mosquitoes give their victim not a moment's peace. One must wear thick clothes, and either don gloves and a screen hat or fight all the time. In camp must be maintained a constant blinding smudge of dead wood of the black mangrove which 'skeets' and men alike detest.... Settlers who pretend any comfort at all screen their houses and keep outside the door a brush of palmetto leaves with which every visitor must beat off the stinging swarm before dodging within. Other settlers keep the smudge pot going and live in smoke. There are also swarms of a terrible great fly, an inch and a quarter in length, whose bite is like a knife thrust, with corresponding flow of blood. No domestic animal except the mule can support life in such a country, and that hardy beast only by being kept in a screened stable and bundled up in burlap when taken out to work.

The famous ornithologist John James Audubon visited in 1832 and while his biographer complained that "mosquitoes were so thick they extinguished his candle," Audubon waxed eloquently, "My heart swelled with uncontrollable delight. The birds which we saw were almost all new to us; their lovely forms ... arrayed in more brilliant apparel than I had ever seen before, and as they fluttered in happy playfulness ... we longed to form a more intimate acquaintance with them."

Today the Everglades National Park brochure states the glades are home to 43 species of mosquito. Nonetheless, at least since this Formative Period the site has tempted humans to survive there.

Shell mounds are to be found between Middle and East Cape, at Little Sable Creek, and near Flamingo. Surveyor Captain Bernard Romans (1775) charted the Keys and coast prior to the American Revolution. In his "directions to navigators" he noted that

> even in this difficult part of the world ... there are always plenty of flamingoes, plovers, and other excellent water fowl to be had. Northwest four miles from this key in Punta Techa, or Sandy Point [Cape Sable]; here was anciently a settlement of Caloosa savages. Tolerable water and excellent venison are to be had here.

Edible wild plants found at the Cape may have added to the ancients' diet. There were ample cocoplums, pigeon plums, the heart and berries of the cabbage palm, the Royal Palm and others.

In 1924 botanist John Kunkle Small wrote that in the area around Cape Sable

> the wild cotton is plentiful in the vicinity of mounds. North of the Florida Keys, where it is widespread, cotton is often found only at and near former settlements of the red-man. He doubtless used it and perhaps cultivated it to some extent. The use of another fiber the white man inherited from the red-man is that of the leaves of the Spanish bayonet for string.

Fragments of pottery uncovered by waves on the wide windy beaches verify Romans' observations of aboriginal presence. Based on the various types of pot shards found at the Cape, archaeologists estimate that people have attempted to survive there on and off for 2,000 years.

Bear Lake Mounds

The National Park Service today operates a lodge and marina called Flamingo adjacent to the Cape. About two and a half miles inland from Flamingo is Bear Lake, the namesake of a habitation site consisting of three mounds. One large mound measures 500 feet by 100 feet by eight and a half feet in height, and the other two are considerably smaller. During dredging of the Homestead Canal in 1922, the south edge of one of the smaller mounds was disturbed by the dredge, uncovering a number of human bones. Thereafter, the canal was diverted around the mound to avoid further site damage.

Sampling of the main mound indicated that probably no more than

30 to 50 people lived there at one time. A number of artifacts were uncovered including pottery shards that established the important dates of habitation. The most interesting feature directly related to the Bear Lake site is Mud Lake Canal. John Kunkle Small writes:

> In addition to the twin mounds back of Flamingo there is an aboriginal canal connecting Mud Lake with the Bay of Florida. During the ages since it was excavated or was in use, it has become largely filled in, but in the shallower places the bottom never becomes wholly dry. This now abandoned channel once made the Cape Sable region an island. By means of it the aborigines could travel from the southern part of the Ten Thousand Islands through Whitewater Bay to Mud Lake. Thereafter they could paddle through their canal to the Bay of Florida without going into the exposed and frequently rough waters of the Gulf of Mexico, a shortcut of about fifty miles.

The canal reportedly measured 20 to 30 feet in width and one to two feet in depth. Considering that it had to have been engineered, then laboriously dug by shells-full or baskets-full of earth, roots and brush chopped away with hand-held shell celts, the canal must have been a major undertaking by many people over a long period of time.

Homestead Site

Hammocks throughout the vastness of the Glades' wetlands are known to have felt man's presence for short durations as he traveled, hunted, fished, and gathered from one location to another. Such sites often yield cultural debitage amounting to disposal or loss of materials by one person or a small band of individuals at a given time. The same sites possibly were used as transient camps over hundreds and hundreds of years. The Homestead Site might be an example of such a transient camp.

The current owner remarked that his property remains high and dry even during times when its surroundings flood, indicating that his current home site, now as during this cultural period, has been a natural hammock within the glades. While digging a very small duck pond at the corner of his garage, the owner uncovered a chert knife, a Levy point, smaller quartz points, and some quartz awl-like tools. The quartz had to have been traded into the South Florida area, possibly as far back as 4,000 years ago based on its association with the Levy point. Or, more likely, the Levy point had been found centuries after it originally had been knapped and was being reused by the finder. The property owner commented that "a lot of small bones were just about on the surface when he first turned the

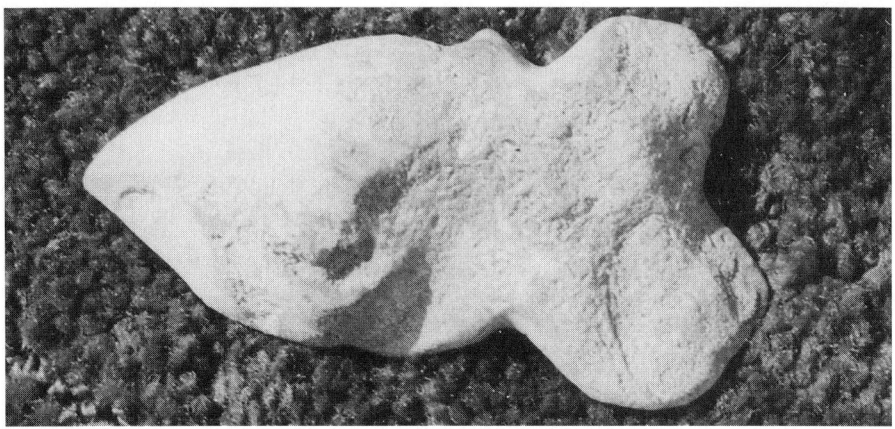

The three and one-half inch long head is unusual. Shaped from local chert, it is too heavy to have served as an atlatl or lance point, but might have proven effective as a ceremonial or war club head.

The Homestead site yields chert and quartz artifacts of diverse origin. The two small quartz points and three quartz hand tools had to have been imported into the area from the north. The large chert knife is of Florida origin. The large Levy point (top) may date back 4,000 years. That does not necessarily mean the site is that old. It might have been picked up later for reuse by the person or people who carried the other tools and points into the area. (Artifacts from the collection of Jim Ford.)

At some time in the distant past, tree leaves and twigs were sealed within a layer of red marl and bits of limestone at the Homestead site. The delicate organics disintegrated, leaving behind permanent impressions in the red earth turned to stone.

earth, and that particular location never had been farmed" (plowed). His interpretation was that the few artifacts had been undisturbed. In any case, the artifacts would appear to constitute a toolkit left behind by an individual at a temporary, dry glades camp.

A bit of additional digging at the Homestead Site revealed pieces of stone on the ancient surface of which are incredibly clear impressions of leaves, stems, and several small coiled terrestrial shells. The stone is hardened red clay from which the area derives its generic name, the Redlands.

This site has yet to be catalogued by Miami–Dade County's Historic Preservation Division. Investigation of the site should be forthcoming in the future. Until then it, and many of the almost 200 other sites throughout the Glades, pose any number of interesting questions.

The Florida Keys

Most long-term habitation sites of this Cultural Area range along the higher elevations of the Atlantic Ridge and, lastly, south through the 130

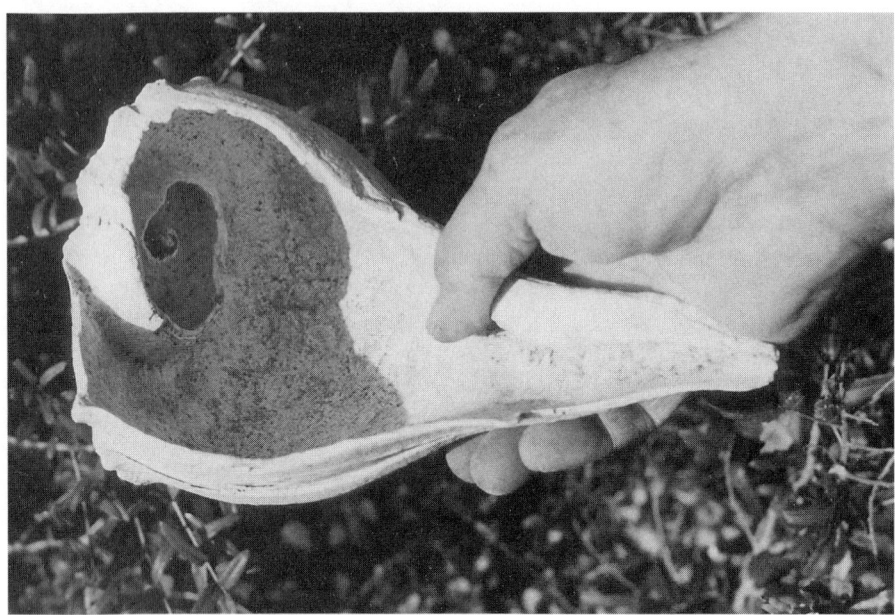

The Redland mound is unique. It contains many large pieces of oolitic limestone as well as black dirt and humus. Lifting one of the rocks revealed several whelk shells hidden in the hollow beneath. Each shell's columella had been removed to create a cup, probably for ceremonial use with caseena, the black drink. Another much larger hollow was lined with shell cups. Obviously, this was a site of great religious or political significance.

mile arc of the Florida Keys. The Keys are considered by some in the field to be the last areas of the Floridan Peninsula subject to human habitation. These islands today average only several feet in elevation above mean sea level, and rarely reach a height of eight or ten feet.

Obviously, in Paleoindian and Archaic times, if man had ventured into this rather inhospitable realm, many of his campsites today would be well below the present shoreline. Freshwater in xeric times certainly would have posed even more difficult problems than during mesic periods. Even in wet times the soils of the Keys are highly porous and water disappears rapidly. Solution holes, potholes, and sloughs containing standing water from rainfall most often would have had to suffice. However, as the underground artesian freshwater pressure increased, some springs were to be found. A few even bubbled to the surface offshore, and potable freshwater actually could be collected out in the ocean, as noted later in many ships' logs.

Consider, too, the effect of storms on islands of such low elevation.

Over the millennia, many storms would have occurred that spawned sufficiently high tides to cover all but the highest elevations on the chain of Keys. Besides their potential as people-killers, the storms probably played a significant role in reworking or completely washing away cultural deposits. Based upon sites found to date, the prehistory of the Keys archaeologically reaches back only to the latter part of the Formative Period. Cultural deposits of earlier man, if any, will have to be uncovered underwater beneath offshore muck, sand, or the living blanket of coral.

Even so, at least 34 Keys sites are recorded to date. Three are located on Key Largo, two of which are middens of black soil, ashes, shell, and bones. One has strata over a foot in thickness of fish bones and strata from one inch to two feet thick of pure ash. The small percentage of midden shells are conchs, whelks, and tulips. The midden is about 75 feet in width and somewhat more than twice that in length with a maximum height of three and one-half feet. The second is only two feet in height and perhaps 200 by 300 feet across, dense vegetation making more accurate measurements impossible at the time of John Goggin's 1940s visit. He was able to collect pottery shards and a few shell tools at the site.

The third is a unique limestone rock mound. This site was popularized in newspaper articles in the 1930s. The rocks are about a foot in diameter and are laid out in rough courses to a height of eight feet covering an area of 100 by 55 feet. A sloping one foot high by 14 feet wide ramp runs east from the mound for 25 feet. A two and a half feet high, by eight feet wide, by 70 feet long limestone ridge is located about 130 feet west of the north end of the mound.

Goggin hypothesizes: "Without a doubt this site was primarily used for ceremonial purposes and may have been of more importance than one would suspect at first glance. The absence of pot shards or other artifacts also tends to indicate that it was of special importance." A similar site, a stone circle 45 feet in diameter and three to four feet wide was reported in 1943 on Boca Chica Key.

Romans (1775) stated that "remains of savage habitations built, or rather piled up of stone" were to be seen in the lower keys on Key Vaca and Key West in his time.

Three mounds have been found on Longboat Key. Four located on Plantation Key were constructed mostly of loose rock, sand and soil. Much of that rich soil has been removed for modern gardens. Pot shards and shell artifacts have been found in rock crevices. An interesting shell and black soil mound on Upper Matecumbe, in addition to shards, yielded large numbers of the small Bleeding Tooth shells, each broken in the same place for extraction of the mollusk for food. A number of ridges dating back

Archaeologist Irving Eyster investigates the Key Largo rock mound. The unusual construction material, its ramps, and configuration cause some in the field to suspect that it served a ceremonial or religious purpose.

between 1,000 to 2,000 years ago extended from the mound into the ocean. Dr. Jack Eaton, visiting archaeologist from the University of Texas, likened the site to many Maya salt works in Guatemala, Belize, and Mexico with which he is most familiar. At high tide, saltwater was trapped between the ridges. Evaporation left the salt behind. Had earliest Keys inhabitants utilized the ridges in that manner, they would have produced a rare commodity probably to be traded to other parts of the country.

On Lower Matecumbe there are several sandy ridges that might have been used for burials that are covered with midden material composed of shell and black soil. Other sites are on Big Pine, Sugarloaf, and Summerland Keys.

Several mounds of questionable origin largely composed of loose rock, sand, and some conch shells are on Plantation Key. Not included is the dark soil typical of middens nor are there pottery shards or other cultural material. If they prove to be ancient in origin they might be ceremonial or burial mounds. Another large mound measuring 200 by 300 feet by six feet in height contained many fish and bird bones, Queen Conch (*Strombus gigas*), Bleeding Tooth (*Nerita peloronta*), and West Indian Top

Shells (*Livona pica*); artifacts were quite scarce.

Goggin reports a "sizeable midden" on Upper Matecumbe with deepest deposits of about four feet. A boat harbor dredged through its side permitted him a clear view of its cross section. He wrote "there does not appear to be any particular stratigraphy and the composition is the usual mixture of soil and shells." He took note, however, of pockets of large numbers of Bleeding Tooth shells, "all broken in the same place for the extraction of the mollusk for food," as was noted above on Plantation Key. He was able to collect pot shards and shell artifacts on the mound surface.

He also reports an isolated burial mound on Lignumvitae Key. He states: "The Indians are reported to have had a great fear of bodies and to have interred them in a constantly guarded place some distance from the village."

The rare bird effigy in Eyster's hand was taken from the Key Largo mound and is felt to further confirm the religious purpose idea. (Rock mound photograph Irving Eyster.)

The burial mound is of coral sand, 50 feet in diameter, and three and a half feet high. Small human bone fragments on its surface indicated its probable use as a burial mound. As seems to be the case with other burial mounds, pottery shards or other artifacts were lacking.

Mounds that formerly existed in Key West have long since disappeared beneath modern construction on a popular crowded island where land is at a premium. A site on adjacent Stock Island was all but destroyed by pot hunters, but has yielded interesting data to the archaeologist's trained eye. Among a variety of pottery shards were shark's teeth, pieces of pumice, animal, bird, turtle, and fish bones, pieces of conch shell (some of which had been worked), a highly polished conch shell columella pendant, bone points, a stone pendant, and a human mandible still holding four of its teeth. A conch shell wall, two or three shells thick, was found in five layers extending down into the mud underwater. This is reminiscent of one on Demorey Key and another described at Marco Island. About two feet inside the wall there was a lot of charcoal and ash amid which was a charred, partial nub of corn or maize. It was only about one and one-half inches long and appeared to be the end of a cob, not enough to serve as proof of agriculture, but as tantalizing evidence in that direction as it was at Fort Center.

It is interesting to note that many of the pottery types found throughout the Keys are those of the southwest Florida coast. Apparently the Calusas and others from the Ten Thousand Islands area down through Cape Sable had little difficulty navigating the very shallow, comparatively calm waters of Florida Bay to reach any area of the Keys. Perhaps they made the trip at that time of year when certain botanicals were ripe for harvest. Seafood always was abundant, as were raccoons and birds. The key to survival would seem to have been to be out of the Keys, back to the relative safety of the mainland before the onset of the time of violent summer storms.

In the early 1500s, Fontaneda wrote accurately of the Keys and people he encountered there:

> There are yet other islands, nearer to the mainland, stretching between west and east, called the Martires; for the reason that many men have suffered on them, and also because certain rocks rise there from beneath the sea, which, at a distance look like men in distress. Indians are on these islands, who are of large size; the women are well proportioned, and have good countenances. On these islands are two Indian towns; in one of them the one town is called Guarugunbe, which in Spanish is Pueblo de Llanto, the town of weeping; the name of the other little town Cuchiyaga, means place where there has been suffering.

Biscayne National Park

An 18 mile long chain of 44 islands lies north of Key Largo; Old Rhodes Key, Totten, Adams, Elliott, and Sands Key to name a few. Together they all comprise just 5 percent of the total of 180,000 acres of Biscayne National Park. The balance is underwater.

An archaeological survey of these Keys revealed 11 habitation sites dating from Glades I to Glades III within the Formative or Ceramic Period. There was possible evidence of pre–Glades habitation, but confirming data to date has proved insufficient. Like the Keys to the south, these islands, too, are based on emergent coralline limestone, upon which is little more than a thin cover of topsoil and sand. What earth and sand there is for thousands of years has been so rearranged by wind and wave erosion that identifying cultural strata at sites is all but impossible.

The Formative or Ceramic Period

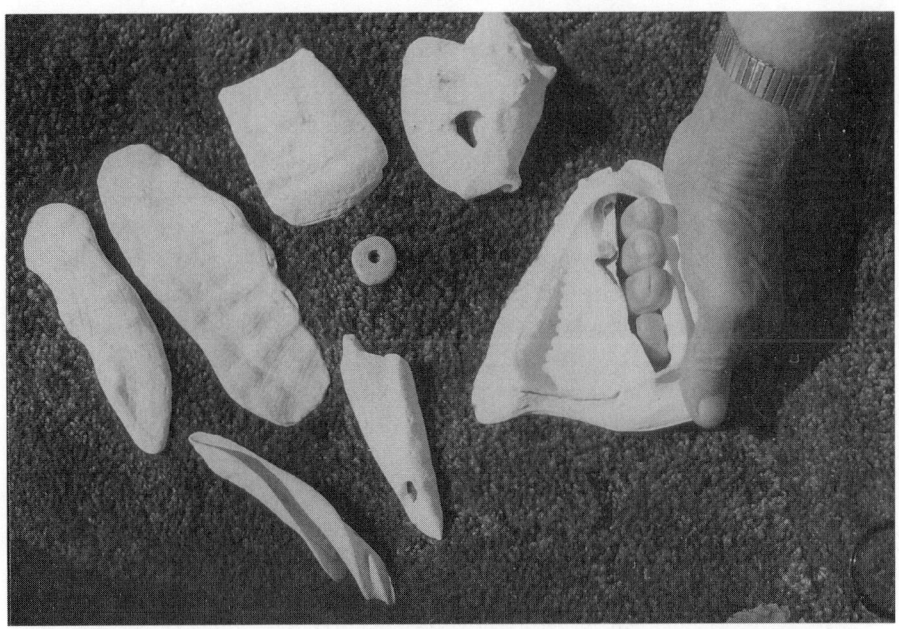

The assortment of shell tools, a bead, and hair pin were found underwater offshore of one of the Keys in Biscayne National Park. The triangular parietal shield of a helmet shell provides a sturdy grip for an effective tool or weapon.

Volcanic pumice has been worn on one side, having been used to hone the edge of various tools and weapons. (Collection of Leonard and Norma Carlson.)

Of interest is the fact that all except one of the sites are located on the ocean side of the islands. Artifacts found included whelk picks, Queen Conch hammers and pounders, celts (both blanks and used), bone pins, shell gouges, columella chisels, and predominantly Glades Plain pottery shards. A few examples of other pottery fragments were found, some having been imported from areas farther south (Matecumbe Incised) as well as north (St. John's Plain) on the peninsula. Middens or mounds typical of the aboriginal people are limited on these islands, there being but one undisturbed shell and midden site on Sands Key that is over 1,000 years old. If more did exist, natural forces, including violent hurricanes, have eliminated all but what little cultural material is found on the surface. Chances are that mounds rarely had a chance to accumulate. Surrounding waters are fairly devoid of oyster beds and there even is a scarcity of other shellfish. Soil and sand are scarce, as well, leaving comparatively little material with which to construct such works.

Faunal remains include mostly broken shell, some fish bones, and turtle bones. This leads one to believe that the sites were visited periodically, rather than inhabited on long–term basis. The people knew that sea turtles nested beginning in early summer (first full moon in June is the old sailor's key), and canoeing to these Keys at the appropriate time yielded a bonanza of large sea turtles and their hundreds of eggs with minimal energy expenditure.

Snapper Creek Site

Several of the site examples discussed have displayed human habitation throughout the entire span of the Formative Period. Obviously, like the Homestead Site, this not always was the case and the Snapper Creek Site in Dade County is another interesting example. One of the Everglades eastward flowing drainages was, and today still is, Snapper Creek. Unlike most streams, it flowed for much of its course underground. The stream then emerged through a sinkhole near the eastern edge of the Atlantic Ridge and made its way above ground to Biscayne Bay. The stream's emergence location served early man as a spring in an area in which freshwater was in limited supply.

A large refuse midden accumulated on top of the unevenly eroded oolitic limestone area surface. The refuse is composed of black dirt and about 10 percent shell and animal bones. The people had ready access to game in the surrounding pine woods, glades, and hammocks. Biscayne Bay was at or very close to its modern level and provided ample seafood, and various native plant foods were abundant and available. Represented

Dark, undecorated, sand-tempered pottery shards were found on the surface of the ground at the Snapper Creek site. While they look of poor quality, they are surprisingly strong and survived perhaps several thousand years as a single shard until a modern house cat knocked them from a table to the floor; hence, five pieces typical of that found throughout the entire Everglades Cultural Area and termed today Everglades plain.

among the shells are whelks, Queen Conchs, cowries, Florida Horse Conch, tulip shells, Venus and other species of clams. Most common among bone remains are turtle, deer, and fish, some small mammals, birds, shark, and very little ray and alligator. Among the rocks nearest the spring, around the old stream bed, and within the holes that pock the limestone surface, pottery shards provide the bulk of artifacts found in the refuse mound. Artifacts other than pottery include shell picks, celts, hammers, a flat sandy limestone grinding stone, and a worked sailfish bill. A recovered shark tooth notched for hafting was the type of tool used to carve wood and bone. Shell and bone tools and points had been sharpened on a worn pottery shard fragment hone, as well as with a number of pieces of volcanic pumice. Several flint chips and fragments of a stone celt indicate travel or trade with more northern people.

The Snapper Creek site probably first was occupied by about 2,000 years ago. At any time during its several hundred years of occupation, the band living there more than likely never exceeded 25 individuals. In most respects the site mirrors those of other small groups living at the same time on the lower east coast south of Miami. Evidence of man-made shelter is lacking, but a friend who had visited the site years ago found sand-tempered pottery shards in rock clefts (which he generously gave me) near what he referred to as "caves" in which he felt the people had lived. Upon close questioning he described rather shallow wave-eroded depressions in the Atlantic Ridge vertical rock face that might have offered some slight shelter from the elements. After several hundred years the site was abandoned and never reoccupied.

Cutler Burial Mound

On the Deering Estate not far from the Paleoindian Cutler Fossil Site (pp. 29–36) is the Cutler Burial Mound. Unlike the former dating back to about 10,000 B.P., the site, thought to be one of the last Tequesta burial mounds left in southeast Florida, is considered to fall within the middle of the Formative or Ceramic Period, from about 1,000 to 500 B.P. The sand mound was excavated in 1876 by Henry E. Perrine, for whom the area of southeast Miami–Dade County is named. He located several skeletons, from each of which the head apparently had been separated prior to interment. All the skulls had been buried face down with the top toward mound center. Seventy-three years later John Goggin investigated the mound, reporting it to be six and a half feet high by 75 feet in diameter. The skeletal materials remain unanalyzed, so the exact chronology and biology of the people remains unknown.

In 1992, category 5 Hurricane Andrew blew through the area and downed a large on-site mastic tree, exposing parts of human skeletons and precipitating erosion. An attempt has been made by Miami–Dade County to stabilize the mound through addition of soil. A path leads from the Deering Estate through the woods to this site. Wooden benches provide a seating area and graphics detail information on this sacred burial mound.

Indian Creek Site

North of the Florida Keys, the islands of Biscayne National Park form the southeastern boundary of Biscayne Bay. John Kunkle Small (1929) writes:

> There was much aboriginal about Bay Biscayne. There were small village sites, kitchen middens, and burial grounds similar to the one just referred to on the eastern side of the bay, even on the islands of Virginia Key and Key Biscayne. The large settlements, however, were on the mainland, especially about the mouths of streams. The larger ones seem to have been near the mouth of Arch Creek, the Miami River, and at Cutler where there were numerous springs of water which come through the subterranean channels from the Everglades.

Small also discusses a village site and burial ground located in the mangroves at the mouth of Indian Creek on today's island of Miami Beach. The remnant Indian Creek once was open to the ocean as a tidal inlet. Small describes the village site as being on a lagoon about a quarter mile from the ocean beach. Its elevation was built up of sand and marine shells as a triangular plateau about 100 yards long on each side. It was elevated well above maximum high water. A burial mound was recorded as being located directly north of the village site. Small feels it "contained a great many bodies, since by reaching under the sand a skull might be located almost anywhere. Many of the skulls were pierced or fractured, indicating death by violence." He feels that the superficial burials might represent the period just after discovery of America. Some of the skulls obviously were not Indian, and most likely were the remains of shipwrecked Europeans who very often were killed the moment they were encountered. If taken prisoner and enslaved, they might have died of disease, or were sacrificed during ceremonies, or at the whim of the casique or shaman. In any event, this site survived into the Historic Period.

Former Florida state archaeologist Vernon Lamme reported a rare burial find north of this area on Lake Worth in what he referred to as a kitchen midden. Truck loads of oyster shells had been removed (along with everything else therein) for construction of the first Dixie Highway. The remaining midden face revealed many strata of ashes, pottery of varying designs, and other midden materials indicative of occupancy by many different peoples. About six feet beneath the surface he and Karl Squires, a former Assistant State Archaeologist of Florida, uncovered "a complete bowl fashioned from the cranium of a human child. The edges were smoothed by some tool and on each side about one half inch from the rim was a neatly drilled hole in which a bail could be fastened (now rotted away, however)." Lamme records this as "first evidence known at that time (1932) of any part of a human skeleton being used for domestic purposes."

Lamme uncovered a second bowl in a burial mound at Chosen, west of Belle Glade, and a good 50 miles west of the first at Lake Worth. It is larger than the first and resides at the Smithsonian where it is identified as "A Trophy Gorget or Bowl of Human Skull." Lamme states the first one as being "remains of Tequesta Indians and the other was evidently of Caloosa Indian craftsmanship."

Arch Creek

Arch Creek is named for South Florida's only natural stone bridge. Obviously, the location was rich in natural resources, and from 1,500[7] to 750 years ago the site north of today's city of Miami supported a sizeable Tequesta Indian population. The site is a high and dry hammock located on the stream that bored through the limestone to form the bridge. The stream provided deep water access to the bay two miles to the east, as well as access to the glades to the west.

Of particular interest is the site's yield of such a rich trove of ceramics that some archaeologists believe the locale supported a vigorous ceramic industry. Further, many large shards were recovered that were almost indistinguishable from one another and display no signs of use. No major clay source has been discovered in the immediate area, but many of the shards and patterns were made of identical pastes.

Other site artifacts include whelk tools and vessels, Queen Conch celts and axes, columella pendants and tools, ground and polished bipointed points of split deer bone, other bone items, and shark vertebrae, of which a few were drilled and others may have been used as ear plugs. A number of pieces of pumice found bore grooves indicating use for polishing and grinding bone and shell materials. An on-site midden is composed largely of conch and oyster shells that had been harvested by Tequestas living there from 500 B.C. to A.D. 1250.

By about 750 years ago,[8] the Arch Creek Tequestas seem to have abandoned the site. Why is subject to conjecture, but Griffin reporters:

> The time period of around A.D. 1,100 to 1,300 [900 to 700 years ago] was a period of climatic change during which some factors not fully understood were apparently at work. We are reminded of the suggested higher sea level around A.D. 1,250 to 1,350 [750 to 650 years ago]. Known sometimes as the Little Climatic Optimum and dated from A.D. 1,000 to 1,200, this was a period of increased warmth which fostered development of Norse colonies in Greenland, brought drought, and initiated cultural

---

[7]*Construction of the Mayan altar with the head of the death god is completed in the city of Copan in Honduras.*

[8]*A commerical and industrial boom begins in northern and central Italian cities. The first golden florins are minted in Florence.*

changes in the Great Plains of North America. At the same time it would have been wetter in South Florida, and we should not discount a possible slight rise in sea level. Even a slightly higher sea level may have forced relocation of sites.

Peace Camp Site

North of the city of Miami in Broward County, near the small town of Davie, west of Flamingo Road and north of Orange Drive, about ten miles from the Atlantic Coast is the Peace Camp Site. It was named from the modern Seminole Indian phrase "Ia monia Okalee," meaning "A place to talk, not fight." Located on what was a peninsula that once jutted out into an old river bed, today the ancient river bed is dry pasture. During very wet times the mound was no more than a small dry hammock above the surrounding water.

Here the author tests the battered ends of a Queen Conch shell that indicate past use as a hammer or pounder by the Tequesta Indians of Arch Creek 1,500 to 500 years ago. Source of the Gastropods was nearby Biscayne Bay, an easy trip by canoe. (Photograph: Todd Zeiller.)

Vertical cross section of the site reveals at least six separate strata, the oldest bottom one dating back to 3,050 years ago (1500 B.C.). The top three strata hold mostly historic artifacts such as cartridge cases, nails, wire, coins, glass beads, and an unusual copper hawk's bell with an iron clapper. Below that strata is an assemblage of ceramics, carved bone objects, fragments of bone awls or chisels, shark's teeth and vertebrae, a drilled human molar, shell beads, *Strombus* celts, *Busycon* tools, *Macrocallista* knives, chert chips, and a stone Broward point or knife. Six shell tools were found in the deepest, oldest pre-ceramic stratum six. *Strombus* celts were made from the thick heavy lip of the Queen Conch shell and, based on the way the edge was honed on pumice or other hard abrasive surfaces, served as an axe, adze, or scraper. *Busycon contrarium* whelk shells were hafted to a stick forced through a hole punched in the outer whorl of the shell and wedged against the shell's columella. The stick was then forced through a hole in the opposite whorl or wedged in a notch cut in the shell's lip. With the tip of the shell's beak (anterior or siphonal canal) sharpened, the tool served admirably as a pick or an efficient war club. Prior to the invention of pottery, whelk and other shells with the columella and inner whorls removed served as dipping, drinking, or cooking utensils. Then the columella, depending on the shape of its tip, served as a hammer, drill, awl, gouge, or wedge for opening oysters and other edible bi-valves. *Macrocallista nimbosa*, the sun ray Venus clam found mostly on the west coast of the peninsula, provided sharp edged shells that served as knives or scrapers.

No human remains were found, which might have been expected since most burial areas would have been located away from the habitation site. The report written about the site concludes that "the Peace Camp Indian lived a full, productive, even artistic life as the small beautiful bone carvings found testify. His home was a well-protected hammock, with food supply abundant from both land animals and the fish from surrounding waters."

Contact with other Indians was evidenced by recovered pottery typical of other areas, and the west coast *Macrocallista* shell tools. In later years, contact with Europeans, probably Spaniards, was indicated by the copper hawk's bell that still rings. Obviously, a fair number of Indians lived at Peace Camp for a very long time.

Madden's Hammock

At the corner of Northwest 154th Street and 87th Avenue in the recently incorporated town of Miami Lakes, on one of the last areas of Miami–Dade County undeveloped land, is 60 acre Madden's Hammock. The site was studied in 1953 and again in 1955 by archaeologists John Goggin and D.D. Laxon. Their work yielded artifacts of three separate cultures dating back 1,000 years or more; Glades culture and Tequesta incised pottery shards and shell beads from about A.D. 1100, majolica Spanish pottery from the A.D. 1500s to 1700, and iron chisels used by Seminoles in the early part of the 20th century. The 1,300 artifacts are curated in the collections of the Florida Museum of Natural History, based at the University of Florida in Gainesville. Scott Mitchell, museum collections manager, says:

> That should tell you how significant the hammock is. Artifacts are the only record we have of some of these historical people. Sites of this nature have disappeared at an alarming rate across Florida, especially South Florida, where development has skyrocketed and natural processes such

as erosion have withered the land. Madden's Hammock is unique because it was used by three different groups.

County archaeologist Bob Carr feels the site dates back at least to 500 B.C., that Tequestas used it as a burial place for thousands of years, and that as recently as the 1800s it was used by the Seminoles as a site for the busk, the Green Corn Dance ritual. Passersby looking north as they drive west along 87th Avenue will see grass-covered mounds just past the hammock's stand of trees. The first smallest mound is separated from the rest by about 50 yards. Then to the west is the highest rounded mound, followed by two rectangular ones of slightly lesser height stretching several hundred yards further west. The round mound and the two rectangular ones appear to be connected by earthen causeways. On the north side, out of sight from the road, are ramps leading from today's ground level to the main mound and the rectangular mounds. A borrow pit parallels this side of the ramp to the rectangular mounds. Standing on top of the main rounded mound, one realizes the immensity of the project undertaken by the ancients to raise themselves above surrounding waters.

The property was scheduled to be developed by its owner family who planned for 85 upscale single family homes. Their application for re-zoning was rejected by the Miami–Dade Commission. In June 2003 the city of Miami Lakes applied to the Florida Communities Trust, the state's land acquisition grant program, for $4.98 million to help purchase the property with the owners' signatures on the trust application as willing sellers. The request was granted and the balance, or 50 percent of the purchase price, will come from Miami–Dade County's Environmentally Endangered Lands Program. The town manager envisions a passive park where boardwalks will allow visitors to enjoy the five and one-half acres of oaks, sweet gums, and palms of the hammock and nature trails, and fishing and canoeing in the park's lakes. The lakes and intermittent wetlands are vestiges of the glades that once covered the area. Until the plan becomes reality, the mounds are merely unusual elevations within a cattle pasture, unrecognized by the passing general public as monuments to an ancient people.

Granada Site

Beneath the present site of the city of Miami was an Indian village named Tequest, as recorded by Hernando D'Escalante Fontaneda (1574):

> Toward the north of the Martires (Keys) and near a place the Indians called Tequest, situate on the bank of a river which extends into the country a distance of fifteen leagues, and issues from another lake of freshwater which is said by some Indians who have traversed it more than I, to be an arm of the Lake of Mayaimi.

Two hundred years later Romans would describe it as "the River Ratones, being a fine stream, and pretty considerable, with a little good rich soil on its banks, where many tropical plants grow."

The river of which he wrote is today's Miami River. Its source basically is the Everglades, beginning in two locations. The north fork is fed by a clear spring-fed lake; the south fork is a drainage from the glades, both emanating from just west of today's 27th Avenue in Miami. Each branch coursed over rapids, or riffs, those on the south branch being lower and less turbulent than those on the north, but both traversable by dugout canoe. South branch rapids were breached years ago and those on the north branch were bypassed by a modern canal. In addition to the northern branch source springs, many other springs were visible along the river's course, bubbling up through the white sand river bottom. The shore of Biscayne Bay obviously was more estuarine than today due to the extensive freshwater input from the springs, the river, and rain runoff. Time and pollution have covered the river's white sandy bottom. Flood control, drainage canals, and heavy freshwater demand have diminished pressure within the subterranean aquifer and, like other springs previously mentioned to the south at Atlantis, Santa Maria, Silver Bluff, and Snapper Creek, they flow no more.

As the salinity of the bay fluctuated, so did the species inhabiting it, and this very diversity of natural habitats provided ample resources for man. Wet prairies and marshes, tropical hammocks, and pinelands were suitable for deer. Gopher tortoises thrived in dry scrub habitats. Fresh waters were abundant with gators, gar, bowfin, sunfishes and bass. The estuarine area of the bay was heavily populated with sea turtles, catfish, snappers, and some sharks. Coral reef habitats ocean side were filled with marine fishes. Whales, dolphins, seals, and manatees frequented the bay and coastal waters. The natural abundance of the area encouraged settlement for thousands of years.

Within the developing city of Miami, a large black earth midden was located at S. E. Second Street and Biscayne Boulevard on the north side of the mouth of the Miami River where it empties into Biscayne Bay. In preparation for construction of the new city's buildings and streets, the mound was leveled. Some 50 or more skeletons were unearthed as the work proceeded, but having little interest in the ancients, the remains were stored

This 1898 photograph was taken of workmen leveling a huge black dirt and sand Tequesta mound in preparation for construction. Located at today's S.E. 2nd Street and Biscayne Blvd., Miami, Biscayne Bay is visible in the background. The mouth of the Miami River and the Granada site are just to the right. (Photograph: Historical Museum of Southern Florida.)

in barrels. During final cleanup of the job site, the barrels were deposited in a pit and covered with sand. The disposal location remains undisclosed to this day.

Other archaeological sites are on the north bank beneath today's James L. Knight International Center, under the parking garage of the DuPont Plaza Hotel and under parking lots north of the DuPont, which was the site of Henry Flagler's Royal Palm Hotel that was built in 1896. Covering an area of nearly five acres between S. E. First and Second Avenues, and between the Miami River and S. E. Fourth Street, it is known as the Granada Site. Two rock mounds, one on either side of the mouth of the Miami River, were nearby. They were reported in 1870 to Boston's Peabody Museum of Archaeology and Ethnology by Jeffries Wyman following his visit a year earlier. However, beyond reporting their existence he apparently left no record of closer investigation of the rock mounds of which no vestiges remain today. The entire area once was a typical hammock bordered on the bay side by mangroves and to landward by pinelands.

The location long ago was described as a very large Indian town, and

The Granada site located adjacent to the Miami River (right) and S.E. 2nd Avenue to the east (above) is under archaeological investigation in this interesting aerial photograph. (Photograph courtesy of the Florida Department of State, Division of Historical Resources, Bureau of Archaeological Research, Tallahassee.)

the Brickell property on the south side of the river was believed to have had a similar village, as evidenced by other black earth mounds on the south bank at the river's mouth, and further west between today's S. E. Second and Miami Avenues. The entire area was referred to by one writer as the "metropolis" of the Tequestas. Occupation of the Granada Site began about 2,000 years ago[9] and continued through the Glades and on into the Historic Period until about 1744.

Archaeological evidence for earlier habitation at the mouth of the river is lacking, but finds along the Atlantic Ridge immediately south of Granada dating back 4,000 years or more would cause one to suspect that perhaps it simply has not yet been discovered. Since buildings, parking lots, roads, and concrete walks cover the entire area, chances of further in-depth investigation are limited, to say the least. Granada Site time frames were established through numerous radiocarbon datings, and the

[9]*London is founded. In Rome Julius Caesar is murdered by conspirators led by Brutus and Cassias.*

ceramic sequence based on well over 13,500 pottery shards recovered from the site's stratigraphic levels.

Artifacts of the Granada lithic assemblage number only 64 of which well over half were made of imported stone. The balance are of local limestone. The eight stone projectile points included therein range from larger knife-like Broward points of Glades I and II Periods to smaller Pinellas points of Glades III times. Other lithic items included a triangular drill, plummet, miniature celt, net weights, anchors, stone disc, mortar and pestles, a pumice smoother, limestone abrader, and limestone sharpener.

About 2,000 artifacts of worked shell were recovered. Of those, more than three-fourths had been shaped from Queen Conch shells, and far lesser numbers of whelks, Horse Conchs (*Pleuroploca gigantea*), and several other species. The Tequestas and their ancestors had utilized the shell materials to make beads, gorgets, pendants, discs, celts, hammers, chisels, gouges, pounders, picks, scrapers, knives, spoons, dippers, receptacles, smoothers, net weights, and anchors.

Over 3,000 specimens constitute Granada's well-preserved, worked bone assemblage, the largest yet found in the southeastern United States. The natural alkalinity of the soil was contributed to by abundant shell refuse and so preserved the vertebrate remains. Vertebrae of sharks, sawfish, and rays were perforated and some held traces of red color. They probably were used as beads, ear spools, labrets, possibly even as spindle whorls for spinning fibers. Shark teeth were drilled, notched, or otherwise altered to serve as hafted knives, drills, or projectile points, and several still held red coloring as pendants or ornaments. Stingray spines generally were modified by partial removal of their natural tiny barbs in preparation for hafting as projectile or fishing spear points. On a few, the barbs were entirely removed and the spines had been smoothed and shaped to a dull ended bipoint. If a shark vertebrae might have served as a spindle whorl, might the bipoints have functioned as weaver's shuttlecocks? Rostral teeth of sawfish showed alteration and wear patterns that suggest hafting and use as various types of pointed tools. Fish bones and spines had been fashioned into pins. One was shaped to resemble a crocodile tooth, and had been drilled for use as an ornament. Barracuda teeth apparently served a number of functions. Halves of barracuda jaws seem to have been hafted for utilization as saws. Pieces of turtle bone, shell, and plastron were found with edges tooled and some drilled. Some may have been parts of rattles. One piece of unidentified reptile bone is flat with engraved surface features, apparently a phallic effigy. A few bits of alligator bone had been honed and polished for hafting as points. Bird long bones also were cut diagonally to form hollow shaft points, or cut square for use as beads.

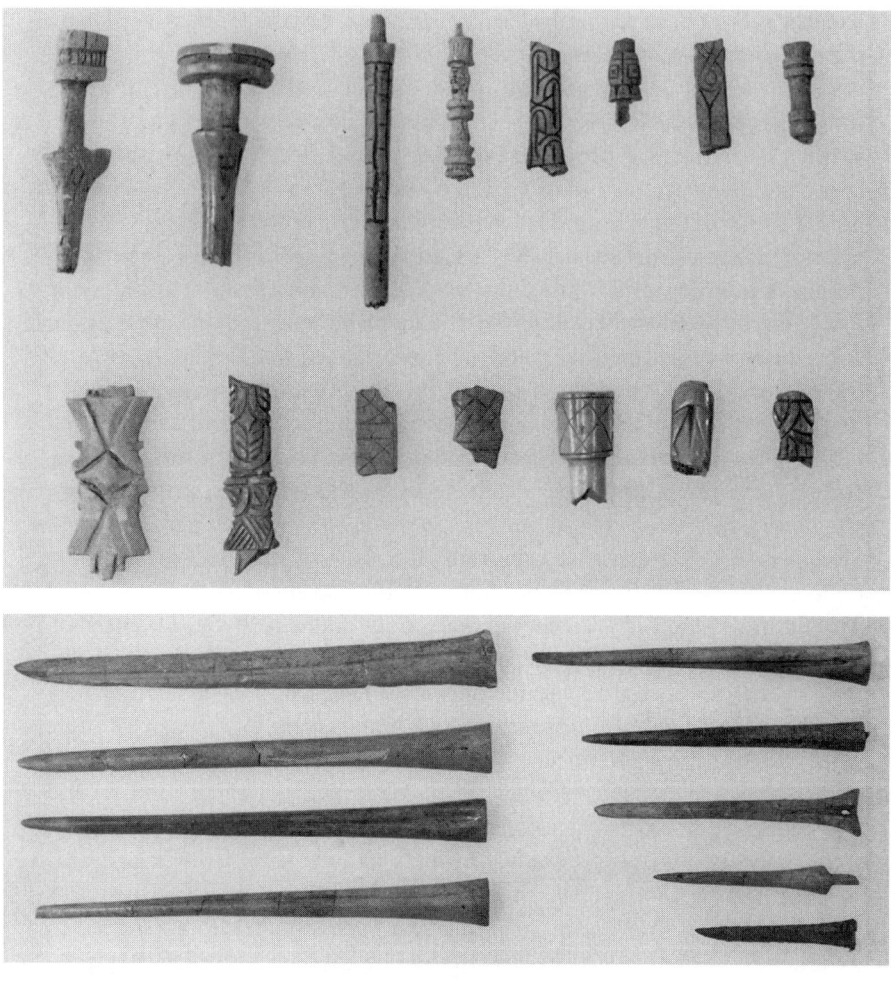

Top: Granada site carved bone artifacts are handsomely decorated hair and utilitarian pins of every imaginable description. Needles, plummets, a gouge, even a few connected fish vertebrae that might have been strung on a necklace, arm, or leg band attest to the artistry of the Tequesta people. *Bottom:* Billfish bills served any number of functions. Note that the base of several smaller ones have been drilled or carved for hafting. (Photographs courtesy of the Florida Department of State, Division of Historical Resources, Bureau of Archaeological Research, Tallahassee.)

Almost three-fourths of the collection consists of tools made from mammal bone, many of which duplicate function of those of shell. There are netting and weaving implements, spatula and gouge-like objects, bipoints, tenoned and socketed points, daggers, awls, pins, and many unidentifiable point, shaft, and head fragments. Many of the bone artifacts had been smoothed and polished during manufacture. Tooled ends were shaped to meet specific need, and some showed definite evidence of having been heavily utilized, then reworked to serve again.

Several plummets were of human teeth, one was of a whale tooth, another from a deer. Other objects were of bone of dog, bear, raccoon, and manatee. The bulk of them had been fashioned from every imaginable bone of deer. A piece of antler displayed many similarities with the reptile bone phallus. Antler had been used for handles, an adze socket, and various implements. A tibia had been crafted into a flute. Phalanges drilled longitudinally were worn as beads. A human foot effigy had been carved from unidentifiable bone. Sufficient bone debitage was found throughout the site to indicate that Granada was a center of a bone tool industry throughout the entire two millennia of the Glades Periods.

The huge amount of faunal material gathered at Granada provided an excellent representative sample for analysis. Many bits of bone or teeth are identified by comparison with modern skeletons. Certain parts of some are extremely durable and distinctive, such as otoliths in catfish, and the unique pharyngeal grinding mill in the redear sunfish. Pieces of turtle shell are easily identified, as are shark's teeth and those of most mammals. Thus armed, researchers were able to determine that within that sample, birds and amphibians accounted for only 1 percent each. Mammals were approximately 6 percent, including the Caribbean monk seal (*Monachus tropicalis*) that became extinct in the 20th century. Sharks, skates, and rays accounted for about 8 percent of the sample, reptiles 11 (predominantly turtle), and bony fish over 74 percent. The percentages clearly indicate that the majority of utilized animal species were aquatic, primarily marine and euryhaline. Fishing obviously was the Tequestas' dominant activity. Mollusks were readily available in the shallow bay. They provided a small percentage of the food biomass and shell for many utilitarian purposes. The Queen Conch comprised about 90 percent of edible molluscan meat. John Griffin (1982) writes:

> There is a dramatic trend toward lower individual size through time for both *Strombus* and *Pleuroplaca*. This is at the same time that these conch are increasing their contribution to the total molluscan biomass. Botkin (1980) calls attention to the fact that predation not only reduces the number of prey available, but also decreases the size of the individuals comprising the prey population.... Inhabitants of the Granada site selectively utilized the molluscan fauna available to them, concentrating on the larger gastropods which could be collected from the bottom of Biscayne Bay. Through time, predation of the dominant *Strombus* species led to diminution in average size of the specimens taken. Decrease in size probably reflects the impact of aboriginal use of the resource.

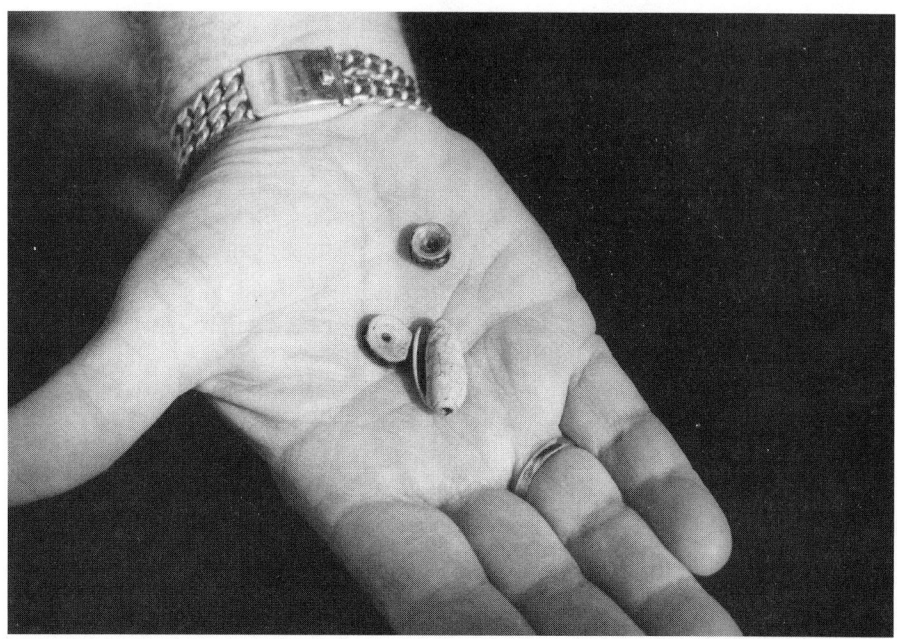

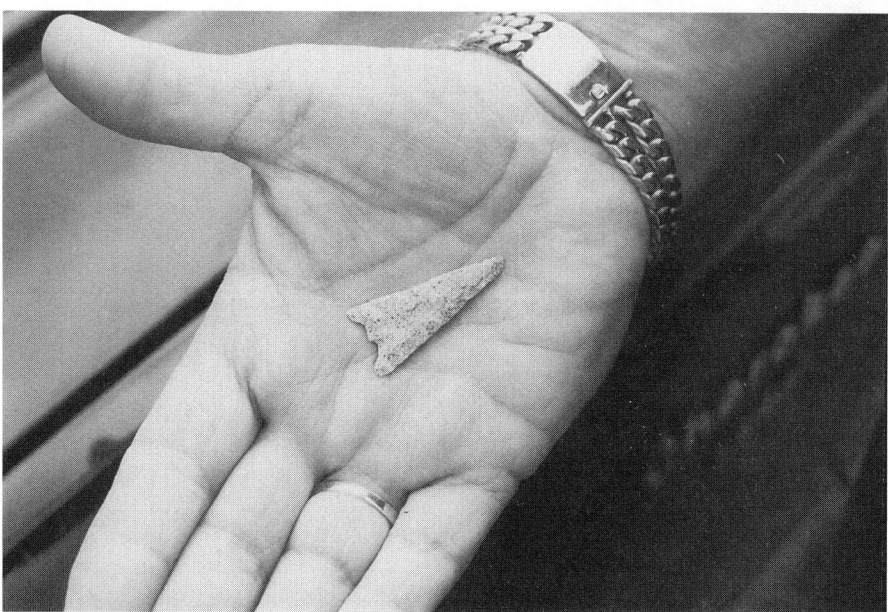

Top: Drilled fish and shark vertebrae and an olive shell were beads left behind at the Honey Hill site by Tequesta people, as was the (*bottom*) prized, small, delicately knapped Hernando point.

Archaeological surveys of Granada indicate that the surface of the site was not covered with an extensive village. Over time, habitations were moved from here to there within its perimeter. It was utilized by the Tequestas as an encampment, from which they traveled seasonally for specific food resources across the bay to islands now known as Virginia Key and Key Biscayne. At other times they may also have traversed the bay to Miami Beach sites, or a relatively short distance northwest on the mainland to the Honey Hill Site located near present Pro Player Stadium. The basis for this theory is consistent use of specific fruits at the different sites. Seeds of mastic, cocoplum, cabbage palm, saw palmetto, sea grape, and hog plum, all trees of tropical hammocks, were abundant in almost all fine screen samples collected at Granada, and most probably were harvested in the fall throughout the years of occupation of the site. Honey Hill was on a hammock on the edge of the glades, less than a day's canoe trip northwest of Granada.

Archaeobotanical analysis there identified the same seeds, but only mastic occurred regularly in quantity. Therefore, the spring and summer abundance of mastic and relative scarcity of other fruits seems to indicate occupation of Honey Hill at that time of year.

Brickell Point Site

Across the Miami River from the Granada Site is Brickell Point, a slightly more than two acre parcel that forms the southeastern shore of the mouth of the Miami River where it empties into Biscayne Bay. From high atop a pedestal on adjacent Brickell Avenue bridge a modern bronze casique, arrow notched in fully-drawn bow, guards his woman and child and their Tequesta home.

The banks of the river at this location were known to be typical hammocks on top of the Atlantic Ridge. Mangroves grew along the bay shore, palms leaned out over the freshwater river, and pine forest to the west stretched to the wetlands known today as the Everglades. In 1871 the pioneer Brickell family built a trading post on this magnificent point. Henry Flagler, whose holdings were on the north side of the river, was feuding with Brickell and by 1903 had covered the south side Brickell site with limestone fill from his dredging of the river. Brickell simply utilized the fill for the foundation of his buildings, and a new home he constructed later in 1909.

Since 1950, a six-building apartment complex completely covered the site. The apartments were razed in 1998. Before the start of construction

Left: The Brickell Point site lies in the shadow of the bridge spanning the Miami River, forever guarded by a bronze Tequesta casique, woman and child. *Right:* A closer look at the bronze statue over the site.

of a developer's twin tower high rise, a six and a half month archaeological investigation of the site was undertaken by Miami–Dade County's Historic Preservation Division. This was made possible by a 1981 ordinance passed by Miami–Dade County and the city of Miami that provided for archaeological investigation of sites to be developed, thought to be the first such ordinance in the nation (see Appendix IV).

Prior to construction of the old apartments, site-leveling apparently had scraped away any pre–1950s historical record, but the dredged fill did protect that which was hidden underneath. Demolition trenches revealed that a black earth midden covered much of the site. Within thin layers of the midden, site supervisor John Ricisak, archaeologists, and a few volunteers uncovered evidence of 1,200 years of pre–Columbian occupation.

Most interesting of all is a fascinating 38 feet in diameter, circular pattern of 24 rather large holes chipped into the limestone bedrock of the Atlantic Ridge. Within the circle there are a few small, natural solution holes, and hundreds of others, man-made, of varying diameters and almost equal depth. Similar small holes have been found farther south along the

The Formative or Ceramic Period

Contract archaeologist John Beriault (right) and Assistant Celestino Diaz wash and search the contents of their shaker screen for artifacts. John paid his assistant the ultimate compliment, saying that Celestino, one day, will discover ancient cities in his native Guatemala.

Atlantic Ridge. They suggest postholes. Ancient tools located on the site, a few shell columellae chisels or picks still upright in the holes, and obvious vertical tool marks confirm the holes man-made origin. Ricisak obtained a chisel made from a fresh Horse Conch columella bound on the end of a stout staff. In no time, he chiseled a hole in the limestone, as had the Tequestas and, possibly, their ancestors before them. Other columellae removed from shells had been pointed to form awls. The thick siphonal canal end of many large shells had been honed to a gouge or chisel-like blade. Columellae pieces had been shaped into plummets. A number of shell celts had been fashioned from the lip of Queen Conchs. Shark teeth were uncovered, some drilled that might have served decoratively, or possibly had been hafted for use as cutting or carving tools. There were finely shaped, double-ended bone points, as well as others shaped from stingray spines. Chunks of volcanic pumice, possibly discharged from Central American volcanoes and floated in on Gulf of Mexico currents, displayed wear patterns indicating use as abrasives to shape, smooth, and sharpen such materials.

Several handball-sized chert nodules were found. Material washed in one volunteer's screen revealed a single worked nodule along with more than a dozen of its associated sharp flakes. Someone long ago had been busy, knapping the nodule into a point or serviceable tool, the sharp flakes saved as microblades. Another sharp-eyed volunteer picked a tiny, one centimeter long chert point from her screen: treasure find of the day.

A few of the shell celts shaped from the lip of Queen Conchs.

The Formative or Ceramic Period 143

Finger impressions just below the rim of this shard are the only reminder of one Tequesta potter who lived at the Brickell Point site.

Three bone points found in the Miami Circle.

A segment of Stingray barb was shaped to an efficient point. Some of the fine barbs are preserved adjacent to the quarter.

A chert nodule that formed within limestone provided material for the tiny 1 cm point.

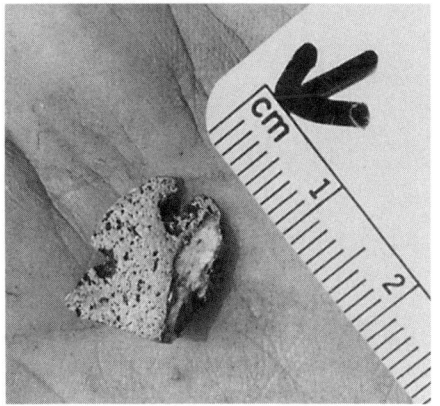

The 1 cm point.

Shell columellae serve as a heavy duty gouge or chisel.

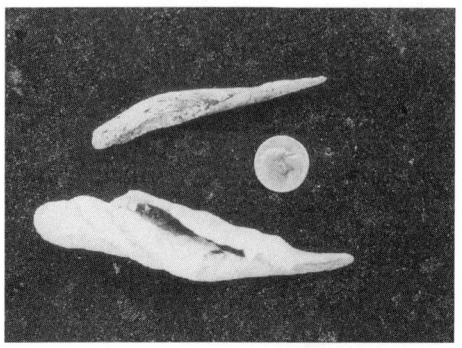

A phid or awl.

Preserved within black midden, soil-filled, man made holes were several shell plummets.

Two hand axes (as well as pieces of two others) proved to have been made of basalt from outcroppings near Macon, Georgia.

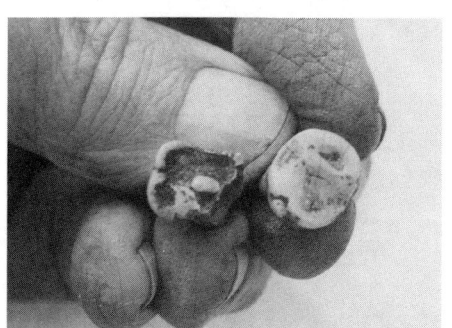

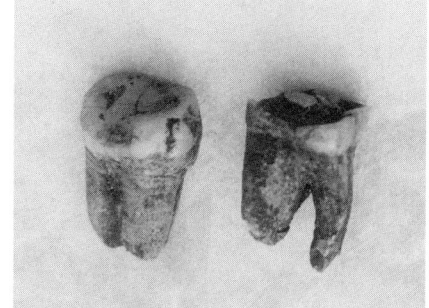

While some Tequesta lost a couple of heavily worn molars, the Brickell Point Circle feature proved not to be a burial ground.

Checked stamped pottery shards found at the circle feature of the Brickell Point site.

Two handsomely polished stone hand axes and broken pieces of several others drew particular scrutiny. Three of the pieces underwent months of physical and chemical analysis by University of Miami geologists. Comparison with a chemical data base of 776 basaltic rock samples from throughout the Americas and Caribbean proved their origin to be basalt outcroppings located close to Macon, Georgia. Apparently, trade had carried them hundreds of miles south to the people at the mouth of the Miami River.

Sand-tempered pottery shards found within the chipped holes display distinctive decorative characteristics that will help confirm later radiocarbon dating. Deptford Simple Stamp pottery, more commonly found in North Florida, was found dating between 500 B.C. and A.D. 500. One shard withdrawn from beneath rocks at the bottom of an arm-length deep chipped hole bore a pattern called Matecumbe Incised; it is said to have been made in the Glades II Period, A.D. 900–1200. The most recent pottery shards found date to about A.D. 1200. Pottery dating to later time is found on the north side of the river, which does coincide with Spanish records that document Indians living only on the north side.

Large numbers of pieces of tortoise and turtle shell and plastron, fish bones, ray and shark vertebrae, seashells, small mammal bones and teeth, and a number from deer indicate what had been cooked in those ancient pots.

Several teeth from the monk seal, which is thought to have been extinct for about 150 years, were found within the circle. Early Spanish records indicate that only the Tequesta elite ate monk seals. A bottlenosed dolphin skull, a complete sea turtle carapace and the articulated vertebrae

The Miami Circle feature at the Brickell Point Site. The rectangular features are remnants of the old apartment building that had been constructed over the site. (Photograph: John Ricisak, Miami–Dade County Office of Community and Economic Development, Historic Preservation Division.)

of a six foot shark, all east-west oriented, a practice often associated with prehistoric human burials, were uncovered in the circle's interior. Cardinal points of the compass, north, east, south, and west, possibly were denoted by holes drilled in the bedrock beyond the circumference of the circle. Various celestial alignments have been proposed. All of these points suggest that the area of the circle was of importance, perhaps the dwelling place of casiques and/or a ceremonial site of shamans. To date, such claims are conjecture.

The archaeobotanical record that might suggest what fruit, seeds, nuts, tubers, and other vegetable matter might have been cooked is scarce, but currently under study, with the exception of a number of carbonized seeds of mastic.

Once the feature was cleared, every single hole cleaned to its bottom, every ounce of soil thoroughly screened, artifacts and ecofacts bagged and coded, Ricisak was lifted in a crane to photograph the circle in its entirety. While work to date hints at rank and ceremony, the unique feature's use, in fact that of the entire site, remains to be determined. While ancient circles are fairly common, none drilled in the rock like this are known in Florida.

Core-sampling throughout the entire two plus acres of the site by state archaeologists, brought in to confirm the site's importance, indicates that about 70 percent of the area contains archaeologically significant features. Charcoal and bone samples have been radiocarbon-dated back to about 2,000 years ago. The land has been purchased from the developer by the State of Florida, Miami–Dade County, and city of Miami to preserve this very special feature and its surroundings for all time. The Florida Secretary of State has appointed an 18 member (including two American Indians) Miami Circle Planning Group to guide the long-term preservation and management of this extraordinary site. The original team of archaeologists and volunteers is anxious to continue their work, for as Ricisak says, "The circle is only a single element on a site that holds so much more."

Immediately north across the Miami River and its shoreline DuPont Plaza Hotel, just east of the Granada site, are three huge parking lots that have covered the six acre area for 50 years. An Indian mound was located only several hundred yards from there (see p. 133) as were old Spanish forts and 19th century Fort Dallas. Henry Flagler's famous Royal Palm Hotel formerly stood at the center of the intersection of the streets dividing the lots. Archaeologists have known more remains of an ancient Tequesta village were hidden beneath the macadam. A development of luxury condominiums, stores, and offices is scheduled to be constructed

Above: Without the bronze plaque generously donated by the Everglades Chapter of the Daughters of the American Revolution, few people would realize that the (*opposite*) 80-foot diameter El Portal mound complex in Miami was a habitation site of Tequesta Indians. (Above photograph: Ed Thompson; opposite photograph: Irving Eyster.)

on the site, but first it must be made available for assessment by archaeologists.

To date this greatly disturbed site has yielded historic artifacts as well as prehistoric materials parallel to those formerly found at Granada and Brickell Point sites; of the latter, a four inch hand ax, edge chipped and worn from use, three inch awl, pot sherds, and hand tools made of heavy shell columellae. A number of postholes, probably for hut supports, have been found near the ancient Biscayne Bay shoreline. Apparently tastes for waterfront property have changed little through the millennia. A small one inch human cranial fragment and four metacarpals also have been uncovered and examined by a physical anthropologist to verify their prehistoric origin. None the less, development of the Miami One complex will progress as agreed to by the Seminole Tribe of Florida's historic preservation officer, with all state law guidelines being scrupulously followed.

Eventually, everything of prehistoric and historic interest found at the site will be placed on display in the One Miami development, but it is far too early to hypothesize as to their possible meaning.

Key Biscayne Sites

Just south of the Granada Site, a strand of offshore islands projects eastward into Biscayne Bay. The largest are known as Virginia Key and Key Biscayne. There was a midden on one of the small islets between the mainland and Virginia Key that was obliterated by the modern connecting Rickenbacker Causeway. Farther out on Key Biscayne there is a black soil midden about 75 feet long by two to three feet high above the high tide line in the mangrove swamp.

During the last century the sea has eroded much of the beautiful beach at the Cape Florida lighthouse on the tip of Key Biscayne. Many years ago, Worth Monroe, well acquainted with this Key, reported an Indian sand burial mound "a short distance north of the old lighthouse." Having been subjected to continuous wave action, even destructive forces of occasional hurricanes, nothing remains of the site that now lies beneath the surface of the semi-tropical sea.

Recently, post–Hurricane Andrew (August 1992) work on the island uncovered three archaeological sites, one of which may prove to be the largest in Miami–Dade County. Some 75 postholes were located in the white sand, possibly providing a picture of what a Tequesta shelter was like. Seeming random placement of the postholes indicate that the dwellings probably were torn down after use, suggesting repeated use of the site. This would be consistent with the hypothesis regarding seasonal use of Granada and Honey Hill sites as well, and Key Biscayne's sandy soil and humus supported more luxuriant flora than the stunted scrub of the Keys to the south.

A brief, easy canoe trip to these islands in early summer would provide access to nesting sea turtles, as it did with the more distant Keys to the south. Fishing would have been excellent. Sharks congregate early in March in large numbers to breed and feed on the apparently inexhaustible numbers of rays that gather as well. At such times, all are easy mark for the aboriginal harpooner. The peoples' adaptation to marine resources is graphically displayed by the fact that, as previously noted, 90 percent of the biomass at Granada was from aquatic animals, with sharks and rays accounting for over half of that mass. Mammalian fauna was more limited than on the mainland, but did include deer and bear. Those and other Mammalia would have added only a small percentage to the larder.

Summer rains provided freshwater on Key Biscayne and Virginia Key, even if wells dug in the sand failed. Should a violent storm or hurricane threaten, nature's warning signs should have provided time to retreat to the nearby relative safety of the mainland. Summer (generally May until early September), too, is the season of mosquito infestation. Life on the offshore keys would have been far more bearable on the breezy ocean side of an island than back near the river and glades. Mosquitoes posed no small problem to man or beast, prehistoric or historic. Marjorie Stoneman Douglas notes in her book *Voice of the River*:

> Mosquitoes were thick in the Everglades.... They seemed to come and go in predictable fashion. The seasons of the Everglades are the mosquito season and the non-mosquito season. During the worst part of the mosquito season, people would move their cows up to Florida City where the cows wouldn't be killed by the bugs. People sent hives of bees down from Pensacola on flatboats to get the mangrove honey, but in the mosquito season, they'd take the bees away so the mosquitoes wouldn't kill them either.

Charles Torrey Simpson (1924) writes of his bout with mosquitoes:

> I can honestly say that for numbers and fierceness as well as for long continued siege what I saw and endured that day exceeded anything I have

ever known before or since. I constantly broke off branches from the scrub which grew along the road and brushed them off as well as I could, but they covered the exposed parts of my body until they were gray, and whenever I wiped them from my face, neck or hands the blood dripped on the ground. The effect of the stings of such a swarm soon became something like that of morphine, producing a stupid, drowsy sensation, and in addition to this my cheeks and eyelids swelled until it was difficult to see. I began to grow weak, my legs tottered and I fully realized that I could last only a limited time.... Twice I went to the side of the track where I dropped down and gave up, but had I remained there I would have been dead in minutes.

D.H. Bellon recorded a rather fanciful account of Everglades insects in *Harper's Weekly*, March 12, 1887:

> Camping in the Everglades was full of discomforts and difficulties. Mosquitoes were ever present and worms by the millions crawled over a daydreamer or sleeper. The trees offered shade and support for a hammock but were inhabited with all manner of worms and insect life with and without legs, bugs, ticks, and other animals, minute in size but terrible in action, which drop willfully and accidentally on the sleeper and promenade with great effect.... Leeches were as numerous in the water as insects in the air.

With little more than smoke of fires and juice of certain plants to serve as defense against such insect hordes, it is no wonder that the aboriginals migrated away from them to the relative comfort and safety of oceanside islands.

The day to day life of the Tequestas and their forebears is difficult, if not impossible, to measure based solely upon the patient efforts of archaeologists. However, their work does permit us to visualize the people, their tools, weapons, foods, and their shelter. Clues to the more esoteric missing puzzle pieces are to be found in the writings of the Spanish, French and English who, almost 500 years ago, ventured into the Floridan Peninsula.

5

The Historic Period: A.D. 1513–Today

Should this search for the people of the southern Floridan Peninsula be concluded at the time of European contact, questions regarding the lifestyle and ultimate disposition of those ancient Native American Floridians as recorded by the Europeans would be unanswered. As mentioned in a previous chapter, human evolution seems to be characterized by long static periods during which there tends to be comparatively little change. In view of these "long static periods," writings of early contact Europeans unwittingly may have offered people of the future glimpses back into prehistory. Those tantalizing views may be confirmed, denied, or altered by the archaeological record. Consequently, it is imperative that this final chapter on the Historic Period follow the foregoing prehistory in effort to provide insight into at least some of these questions. This is particularly true in regard to the chiefdoms. None have existed for the past 200 years bearing the names of Ais, Jeaga, Mayaimi, Boca Ratones, Calusa, or Tequesta to name a few. Where did they go? What happened to them? Are they the honored ancestors of Florida's Native Americans of today, the Seminoles and Miccosukees?

Perhaps Robertson regarded the descriptive writings of Florida by Ponce de Leon, Panfilo de Narvaez, Hernando de Soto, Tristan de Luna, and others in a most enlightened manner when he wrote in 1777:

> The Spaniards who first visited America, and who had an opportunity of beholding its various tribes, while entire and unsubdued, were far from possessing the qualities requisite for observing the striking spectacle presented to their view. Neither the age in which they lived, nor the nation to which they belonged, had made such progress in true science, as inspires enlarged and liberal sentiments. The conquerors of the New World were mostly illiterate adventurers, destitute of all the ideas which should have directed them in contemplating objects, so extremely different from those

with which they were acquainted. Surrounded continually with danger, or struggling with hardships, they had little leisure, and less capacity for any speculative inquiry. Eager to take possession of a country of such vast extent and opulence, and happy in finding it occupied by inhabitants so incapable to defend it, they hastily pronounced them to be a wretched order of men, formed merely for servitude, and were more employed in computing the profits of their labor, than inquiring into the operations of their minds, or the reasons of their customs and institutions.

Not only the incapacity, but the prejudices of the Spaniards, rendered their accounts of the people of America extremely defective. Soon after they planted colonies in their new conquests, a difference of opinion arose with respect to the treatment of natives. One party, solicitous to render their servitude perpetual, represented them as a brutish, obstinate race, incapable either of acquiring religious knowledge, or of being trained to the functions of social life. The other, full of pious concern for their conversion, contended that, though rude and ignorant, they were gentle, affectionate, docile, and by proper instructions and regulations might be formed gradually into good Christians and useful citizens. This controversy was carried on with all the warmth which is natural, when attention to interest on the one hand, and religious zeal on the other animated the dispute. Most of the laity espoused the former opinion; all the ecclesiastics were advocates for the latter; and we shall uniformly find that, accordingly as an author belonged to either of these parties, he is apt to magnify the virtues or aggravate the defects of the Americans beyond measure. Those repugnant accounts increase the difficulty of obtaining perfect knowledge of their characters, and render it necessary to peruse all the descriptions of these by Spanish writers with distrust, and to receive their information with some allowances.

Several centuries later Professor Dean Snow, Chairman of the Department of Anthropology, State University of New York, concurred with Robertson's evaluation, as do many others today. In 1976 he wrote of the early Spaniards: "They were adventurers looking for means to make themselves rich, famous, or powerful, according to a rather specific set of European values."

With that in mind, it was recorded that the first voyage sanctioned by the Spanish crown to this new land was undertaken by Juan Ponce de Leon, former Spanish governor of Puerto Rico. His patent and travels were recorded by one Antonio de Herrera, royal historiographer of Spain. Previously he had completed the 1493 second New World voyage with Columbus. Thereafter, he had been appointed Adelantado of the western section of the West Indian island of Hispaniola that is today's Republic of Haiti. Ponce was anxious to and received permission from the King of Spain to explore regions west of the Indies. He sailed along with two other ships on March 3, 1513, on his flagship San Cristobal from Puerto Rico, passing the island of Abaco on the 27th and reaching the Floridan Peninsula on the 28th. He landed on the northeastern coast of the peninsula near St. Augustine on April 3rd, 1513, and claimed for Spain all territory east of Mexico, including today's North Carolina all the way south through the Florida Keys; this vast area he referred to as "La Florida." Within the week, he sailed south along the coast. He first sailed into Biscayne Bay (today's Miami area) on the southeast coast on July 3rd of that same year and encountered the people living there.

There is evidence of prior contact between Spaniards and aboriginals of the southern part of the Floridan Peninsula. It was noted that at first contact with the west coast Calusas, the Spaniards encountered an Indian who was conversant in Spanish. Some historians feel the south Florida aboriginals encountered the Spanish language as they traveled in their large dugout canoes and catamarans along the coasts of the peninsula, and the Gulf of Mexico southward through Mesoamerica, as well as to the Bahamas and isles of the Caribbean. They were perfectly capable of such trips, as it was recorded that they accomplished them in their huge canoes that held 40 men or more. Two such hulls lashed together catamaran-style easily might hold twice that number, and would be far more stable at sea. It is appropriate to think of them using catamarans, since a similar toy double canoe was found by Cushing among the archaeological treasures of Key Marco.

Of course, by then natives more than likely traveled in both directions. The South Florida Indians surely would have welcomed the friendly Arawak traders. Perhaps it was Arawaks who survived earlier contact with Spaniards from Hispaniola, Puerto Rico, and Cuba who brought the strange new language to Florida. The more peaceful, rather helpless Arawak Indians of the Caribbean Islands already had been enslaved by the Spaniards within a few years of discovery by Columbus. They had rapidly died of cruelty, exhaustion, and European diseases to which they had no immunity. Of the latter, Europeans carried with them cholera, typhoid fever, scarlet fever, measles, bubonic plague, pleurisy, diphtheria, whooping cough, mumps, gonorrhea, chancroid, syphilis, typhus, influenza and other respiratory diseases. Or was it Spaniards whose visits prior to 1513 went unrecorded, who offered only three choices to any indigenous people they could find: pay tribute, work, or become enslaved. Those Spaniards were hunting for slaves to work their plantations on the islands to the south.

By 1502 the first slaves from Africa had been brought to the Caribbean islands to serve in place of the exhausted supply of native Indians. These unfortunates from that continent introduced the anthropod-borne diseases malaria, yellow fever, and probably dysentary to their European captors as well as to the Indians. As the availability of ready manpower diminished, the Spaniards also ranged to the northwest for their slaves. Several archival letters concerning a 1520[1] cruise under license of Lucas Vasquez de Ayllon document this fact. One states that his caravel encountered another that was returning from an unsuccessful cruise among the Bahamas for Caribs, the object of the expedition being to capture Indians in order to sell them as slaves. The ships sailed together and, in due course, they landed some 20 of their men ashore. They soon were surrounded by Indians whose friendship was won with a few simple gifts. Days later, the Spaniards, ignoring the orders of Ayllon to cultivate friendly relations with Indians, seized some 70 of them and sailed away.

A second letter from a Peter Martyr is more descriptive and reads in part:

> These Spaniards visited all the Lucayas but without finding the plunder, for their neighbors had already explored the archipelago and systematically depopulated it.... They were driven by a sudden tempest which lasted two days, to within sight of a lofty promontory. When they landed on this coast the natives ... at first rushed in crowds to the beach, eager to see ... the natives fled.... Our compatriots pursued them ... and captured a man and women.... They took them to their ships and after giving them clothing released them. Touched by this generosity, serried masses of natives again appeared on the beach. When the sovereign heard of this generosity, and beheld for the first time these unknown and precious garments ... for they wear only skins of lions and other wild beasts ... he sent fifty of his servants to the Spaniards, carrying such provisions as they eat. When the Spaniards landed, he received them respectfully and cordially, and when they exhibited a wish to visit the neighborhood, he provided them with guide and an escort.... The Spaniards ended by violating this hospitality. For when they had finished their explorations, they enticed numerous natives by lies and tricks to visit their ships, and when the vessels were quickly crowded with men and women they raised the anchor, set sail, and carried these despairing unfortunates into slavery. By such means they sowed hatred and warfare throughout the peaceful and friendly region, separating children from their parents and wives from their husbands.... These people are white and larger than the generality of men. When they landed some of them searched among the rubbish heaps along the town ditches for decaying bodies of dogs and asses with which to satisfy their hunger. Most of them died of misery, while those who survived were divided among the colonists of Hispaniola, who disposed of them as they pleased, either in their houses, the gold mines, or their fields.

---

[1] *By 1521 Cortez had occupied Tenochtitlan and was sacking and destroying the capital of the Aztecs. The Conquistadors established Mexico City, their own capital, on the site.*

Voyagers from the Caribbean who preceded the Spaniards were not always welcomed by people of the Floridan Peninsula. Among them were the aforementioned Caribs, fierce and warlike aboriginal people who were described by the Europeans as "delighting in blood," and who were said "to devour the flesh of those unfortunate enough to become their prisoners." All who encountered them, Arawaks, Calusas, Tequestas, everyone, feared these supposed cannibals. It is little wonder, then, that initial recorded contacts of the peninsula's people with the Spaniards were met with fear. The Conquistadors must have seemed no different to them than the fierce Caribs. The strange men, armored and armed with fearful weapons, killed them on whim, took prisoners for enslavement and carried them away, never to be seen again. Many of the indigenous people surely believed that their unfortunate fellows had been eaten.

Men who had been shipwrecked along the Florida peninsular shores most often were slaughtered immediately. A very few, like the previously quoted Fontaneda, were enslaved. Their lot was tenuous; a slave might be killed at any time, being subject to the whim of the "cacique," a word for head person or chief that the Spaniards adopted from the Arawak language spoken by the Taino tribe of Hispaniola. At least one slave was selected to be sacrificed for any of a number of celebrations each year. To demonstrate his own power, two castaways who had been sent as tribute to the casique Tequesta by the cacique Ays (Ais) were promptly killed. On occasion a captive Spaniard was ransomed; in one instance a Captain was kept alive after eight of his crew had been lulled by offers of friendship, then killed with darts (atlatls) and clubs. It must be noted that not only Europeans were enslaved by the indigenous people of La Florida. Like the Europeans, the natives did not hesitate to impress anyone capable of bearing their burdens.

In just 20-odd years from the time Columbus set foot in the New World, Old World conquerors reached "La Florida." With the exception of the Jesuit Brothers, whose intent was to convert the heathens to Christianity, their goals were economic.

On September 26, 1514, a delighted King Charles V issued Ponce a new patent to colonize Florida. He embarked from Puerto Rico on February 20, 1521, with two ships loaded with colonists, cattle, and seeds for crops. His destination was Big Mound Key. Within the week he sailed south along the peninsula's coast. Royal historiographer Herrera recorded location names along the way, some Indian, some given by Ponce or his crew; Cequescha, (later termed Tequesta) a village on Biscayne Bay, the Martires (the Keys), and Achecambei (Matecumbe) to name a few. Ponce traveled around the Keys, then north as far as Charlotte Harbor and Big

Mound Key, near present Ft. Meyers. It was here at what he felt was their capitol he encountered the Calusas of the southwest coast.

The Calusas appeared to dominate other south Florida groups for a long period of time, during which they paid tribute to their cacique Carlos. Many of their mound-bearing sites have been discussed in earlier chapters. Garcilaso de La Vega (son of a Peruvian Inca princess and a Spanish nobleman known as *The Inca*), a member of an early Spanish expedition, transcribed a description of the use of those mounds in a Calusa village. It might be typical of their villages at that time, in that locale, which is supposed to have been somewhere in either today's Manatee or Sarasota counties.

> As many as possible, at least the chief and under lords, had houses built on the flat of the top mound site, according to the grandeur of the ruler, accommodating from ten to twenty houses, for the dwellings of the family and the serving people. On the flat at the foot of the hill [mound] they make a quadrangle square, according to the size of the village which is to be located around it.

Their society was described as being matrilineal. While a cacique married his sister, thereby perpetuating the family leadership lineage, other men moved to the family of their women even if they were of different groups. The practice would seem to have caused a constant state of flux among the numbers of men within a given group. Of course, other major factors entered into the equation, such as conflicts or food availability. For whatever the reasons, the dominance, even existence of any band or chiefdom had to have fluctuated over time.

Tequestas dominated the east coast, but still were somewhat under political control of Carlos, casique of the Calusas. Later, Jonathan Dickinson noted that the Ais were dominant on the southeastern Atlantic coast. Later still the indigenous population seems to have broken into numerous smaller groups.

Historical research in the colonial archives in Spain and Cuba reveal eyewitness accounts of 1566 ceremonies in which the Calusa king (casique) allied with the Spanish governor in a building that held 2,000 standing people. The Spaniards recounted witnessing rituals inside great temples decorated with carved and painted wooden masks. The casique was said to have "power of life and death over his subjects and was thought by them to be able to intercede with the spirits that sustained the environment's bounty. Commoners supported the nobility and provided them with food and other material necessities."

Ponce's colonists survived in the kingdom of the Calusas only a few months. During his first voyage the Indians had been friendly, even helpful. On a second voyage he found that in his absence the friendly Indians he had known had been killed and the new generation hated and feared the white men. Many of the colonists were ill or wounded. Attacked by the Indians, Ponce suffered a grievous arrow wound, so all returned to Cuba where within a few months he died.

Several years later during the 1528 Florida expedition of Panfilo de Narvaez, the Spaniards, lacking knowledge to sustain themselves in so abundant an environment and near starvation, were reduced to eating several of their dead companions. This, among other things, served to reinforce the horror and indignation of the Indians against the Spanish. Conversely, a number of Spanish and, later, French chronicles repeat charges that the Indians ate the flesh of their own dead as well as that of their captives. However, those stories seem to have served more to justify the Europeans' behavior, and lay blame based on Christian bias upon the "heathen savages." Narvaez left absolutely no doubt regarding his attitude toward the natives:

> With the aid of God and my own sword, I shall march upon you; with all means and from all sides , I shall war against you; I shall compel you to obey the Holy Church and his Majesty; I shall seize you.... Your property I shall take and destroy and every possible harm shall I work you as refractory subjects.

His arrogance yielded little more than total disaster and death for him and most of his troops at the hands of his "subjects."

In spite of politics, misconceptions, and terrible treatment by all sides, the Spaniards persisted. An early reference to Matecumbe in the Keys was in a 1573 petition from Pedro Menendez de Aviles, founder of St. Augustine in 1565, to the Spanish crown. The petition was accompanied by affidavits from other Spaniards attesting to incidents of cruelty, murder, and enslavement perpetrated by the Matecumbes. Menendez stated that the Indians of South Florida were a bloodthirsty menace to the Spanish, particularly shipwreck survivors, and he requested permission to exterminate or enslave them. Wisely, in 1574 the crown rejected his petition for "giving up as slaves the Indians of Florida."

Menendez landed at Tequesta on Biscayne Bay in 1567 while on his return to Havana. His contract with Philip II required him to cause the indigenous people to become loyal, obedient subjects of the crown by their conversion to Catholicism. Therefore, he established Jesuit brothers Rogel and Villareal, along with a small body of soldiers, in a fort on the shore of Biscayne Bay. They seem to have been accepted because of the gifts of food, metal tools, and weapons they brought to the people.

A letter dated January 29, 1568, from Francisco Villareal to Rogel described some aspects of Tequestan daily life:

> I and all of us here remain in good health, glory to God who helps us to endure in this land trials which would appear insufferable in another place. I say this for we have had for the past three months or more a plague of mosquitoes so bad that I spent several days and nights without being able to sleep an hour. On top of this we suffered some days for lack of food. I say no more about this but to add that the only sleep we could obtain was close to the fire and half smothered in smoke, otherwise one could not endure it. At this time the majority of the Indians went to an island a league from here to eat coconuts and palm grapes. No more than thirty remained here.... There are many here now because some of the nearby villages have come to help in building a house for the chief. They now have food from the whales they kill and from fish. Before they suffered from hunger for two or three months so that they failed to attend because they all said they were hungry and begged that which I had little to give them. With all this the young Chief is very fond of the Christians and it seems he will become one.

Brother Villareal continued with a description of his reciting the evangels and making the sign of the cross over a sick child, who then became well within a day. Similar successes followed. In one case, the child failed to improve and when death seemed imminent

> The parents called the witch doctors in who performed all sorts of rites squeezing her body till it seemed they would crush her, but as she continued getting worse the witch doctors said they might have cured her if I had not touched her.... I baptized her on the ninth and she died on the eleventh of January. The interpreter told me that if we had not been present, according to their old law they would have sacrificed four other children with her.

Lopez de Velasco wrote in 1569 that when a cacique of Tequesta died, his body was disjointed and the larger bones from his body were placed in a box. This was placed in his house where every villager went to see them, believing the bones to be their god. The cacique's retainers were sacrificed, presumably to serve their master in the hereafter. If a child of the cacique died, other children were sacrificed.

A study of documents in the Archive of the Indies in Seville, Spain revealed that a Calusa chief from the Pine Island Sound area visited Havana in July of 1692. He requested establishment of a mission in his kingdom. Four Franciscan missionaries arrived five years later to fulfill his request, bringing with them sufficient clothing and food for their personal use for six months.

The Calusas became hostile to them when they realized the missionaries' stores were not gifts from the King of Spain for them. The missionaries compounded their own problems by violating the sanctity of the Calusas' religious temple. It was described as "long, wide and tall, with one door. Within the building was a mound or alter of earth with a kind of bench or lattice work structure on its peak. The alter and walls were covered with matting and grotesque long-nosed masks." They entered therein reciting the Holy Christian rosary and holding aloft an image of the Virgin Mary.

The Calusas had had enough. They placed the missionaries and their few belongings in a canoe and sent them south toward Havana. However, on their way south other Indians relieved them of their provisions and stripped them naked. The hapless Franciscans, who were probably fortunate to have had their lives spared, arrived naked and starving in Havana 30 days later in February of 1698. It is apparent that most of the Calusas had little interest in being converted to Christianity; they were interested only in whatever food, clothing, tools, or weapons the Spaniards might provide.

Jerald Milanich, Curator and Program Director of Archaeology, Florida Museum of Natural History, University of Florida, Gainesville (1996) explains the role of the missions in Spanish Florida: "Now more than ever we see that missions and colonization were integrally related. Christianized Indians enabled a colony to function. In return for providing religious education for native people, the Spanish could harness them as workers in support of colonial interests. Religious instruction made them obedient, productive members of Spain's empire." He continues, describing their extraordinary piety and intensity of devotion, even in the eyes of the friars.

Nonetheless, as hard as they tried, the Jesuits soon gave up their efforts in south Florida for the more fertile fields of New Spain or Mexico. But, from their efforts and those of Franciscan missionaries who followed, a strong political bond eventually did develop between the Spanish and the Indians.

The name of both the cacique and his village at the mouth of the Miami River was understood by the Spaniards to be Chequescha, later to be written and pronounced by them as Tequesta. Tequesta is the nomenclature that has survived ever since. The parallel nomenclature of a chief and his village was not an aboriginal system. Rather, the Spaniards simply were following their own feudal custom then prevalent in Europe. Tequesta is, in fact, Ponce's first recorded place name in southeast Florida.

The Tequestas (variant spellings of which are Tekestan or Tekest) were recorded as being hierarchically organized, fisher-foragers who moved periodically to take advantage of seasonally abundant resources. This has been confirmed in the aforementioned archaeological records of the Granada, Honey Hill, and Key Biscayne sites.

In 1598 the Spanish governor of Florida wrote that "the natives had no settled habitations, as they did not grow maize, but wandered about in search of fish and roots. Their chief vegetable diet consisted of palm berries, cocoplums, and sea grapes. There were animals in abundance, such as deer and bear, and fish as plenty as they please and alligators for food."

Many Europeans probably were unable to recognize more than a limited number, if any, of the botanical resources at hand. As Dean Snow writes: "They were adventurers looking for means to make themselves rich, famous, or powerful, according to a rather specific set of European values." They had little or no knowledge of botany and could starve to death standing in fields of plenty.

The indigenous people had learned to take full advantage of their environment and did utilize a broad spectrum of botanicals. While it has been noted previously that vegetable matter leaves little behind in the way of artifactual representation, acorns, hickory and other hard nuts, utilitarian wooden objects, carvings, and some twisted and woven fiber materials may be preserved in peat or other anaerobic deposits to survive the millennia. Plant stems, leaves, fruit, blossoms and other soft parts often are lost forever in tropical and semi-tropical environments. Seldom are they found on the lower Florida Peninsula even in fossil form. However, carbonized bits of woody trees and plants, seeds or fragments thereof, and pollen under certain conditions may survive and provide plant identification within given time frames based upon the context within which they are found.

Listed below are many, but obviously not all, of the plants that were to be found throughout the entire 12,000 years of prehistory discussed within this text. Environmental changes from xeric to mesic conditions and the reverse; even changes in sea level altered the botanical assemblage accordingly. While archaeological research seems to indicate that botanicals may have constituted a small percentage of the aboriginal diet, the following substantial list might indicate that hunger need not have plagued early man. Edible botanicals other than those found in references for this work probably were available as well, and may be found in Dr. Julia Morton's *Wild Plants for Survival in South Florida*.

Plant	Scientific Name	Harvest Time	Edible Parts
Live Oak	Quercus virginiana	Aug–Dec	acorns
Cow Basket Oak	Q. michauxii	"	"
Chinquapin/chestnut	Castanea spp.		nuts
Pignut Hickory	Carya glabra	fall	nuts
Hog Plum	Ximenia americana	Mar–May, Sep–Nov	fruit
Saffron Plum	Bumelia celastrina	all yr.	fruit
Seven Year Apple	Casasia clusiifolia	all yr.	fruit
Cocoplum	Chrysobalanus icaco	Jun–Nov	fruit and kernel
Satinleaf	Chrysophyllum oliviforme	spring	fruit
Gopher Apple	Licania michauxii	all yr.	fruit
Pigeon Plum	Coccoloba diversifolia	Sep–Nov	fruit
Sea Grape	C. uvifera	Sep–Nov	fruit
Geiger Tree	Cordia sebestena	Mar–Nov	fruit
Bay Tree	Persea borbonia	all yr.	leaves
Red Mangrove	Rhizophora mangle	all yr.	peeled sprouts
Black Ironwood	Krugiodendron ferreum	Aug–Nov	fruit
Mastic	Mastichodendron foetidissimum	all yr.	fruit
Paradise Tree	Simarouba glauca	Apr–May	fruit
Black Gum	Nyssa sylvatica		fruit
Wild Cherry	Prunus myrtifolia		fruit
Ground Cherry	Physalis spp.		fruit
Wild Persimmon	Diospyros virginiana	Sep–Nov	fruit
Red Mulberry	Morus rubra	Jun–Aug	fruit
Wild Fig	Ficus citrifolia	all yr.	fruit
Silver Palm	Coccothrinax argentata	Sep–Dec	fruit
Cabbage Palm	Sabal palmetto	Sep-Dec	fruit, heart and bud
Saw Palmetto	Sereona repens	Sep–Nov	fruit and bud
Custard Apple	Annona reticulata		fruit
Pond Apple	A. glabra	Jun–Nov	fruit
Wild Papaya	Carica papaya	Feb–Nov	fruit
Wild Lime	Zanthoxylum fagara	Jun–Oct	fruit
Guava	Psidium guajava		fruit
Long-stalk Stopper	P. longipes		fruit
Wild or Mud Potato or Groundnut	Apios americana	all yr.	tuber
Wild Sweet Potato	Ipomoea spp.		tuber and flower
Moonvine	Ipomoea spp.		tuber and flower
Coontie (Arrowroot)	Zamia intergrifolia	all yr.	tuber
Smilax (greenbriar)	Smilax spp.		tuber and shoots
Cattail	Typha spp.		tuber and stem heart
Spike Rush	Eleocharis spp.		tuber

Plant	Scientific Name	Harvest Time	Edible Parts
Sagittaria	Sagittaria latifolia		tuber
Nut Grass (Nutsedge)	Cyperus spp.		tuber
Wild Tomato	Lycopersicon esculentum		fruit
Gourd	Lagenaria siceraria		seeds
Squash	Cucurbita pepo		seeds
Indian Pumpkin	C. okeechobeensis		seeds
Marlberry	Ardisia escallonioides	Mar–Apr	fruit
Elderberry	Sambucus canadensis	all yr.	fruit and flowers
Sugarberry	Celtis laevigata	Apr–June	fruit
Blackberry	Rubus spp.		fruit
Huckleberry	Gaylussacia spp.	May–June	fruit
Dwarf Huckleberry	Lasiococcus dumosus		fruit
Locustberry	Byrsonima cuneata	all yr.	fruit
Beautyberry	Callicarpa americana	Sep–Nov	fruit
Muscadine Grapes	Vitus rotundifolia	Aug–Oct	fruit
Fox Grape	V. cordifolia		fruit
Passion Fruit	Passiflora foetida		fruit
Blueberry	Vaccinium myrsinites		fruit
Sunflower	Helianthus annuus		seeds
Prickly Pear	Opuntia spp.	all yr.	fruit and juice
Fiddlewood	Citharexylum fruticosum	all yr.	fruit
Goosefoot (Pigweed)	Chenopodium spp.		seeds and leaves
Portulaca	Portulaca spp.		leaves
Purselane	P. oleracea	all yr.	leaves
Sourdock	Rumex patientia		leaves and juice
Sawgrass	Cladium jamaicensis		buds
Coral Bean	Erythrina herbacea	spring	leaves
Meadow Beauty	Rhexia virginica		fruit
Spanish Bayonet	Yucca aloifolia		fruit and flowers
Yaupon Holly	Ilex vomitoria		leaves for black drink
Stopper	Eugenia axillaris	all yr.	
Saltwort	Batis maritima	all yr.	leaves-seasoning
Sea Purslane	Trianthema portulacastrum	all yr.	leaves and stem
Sea Blite	Suaeda linearis		leaves-seasoning
Watershield	Brasenia schreberi		leaves and roots
Spatterdock	Nuphar lutea		seeds and roots
Bulrush	Scirpus spp.		seeds, shoots, tubers
May Pop	Passiflora incarnata		fruit
Amaranth	Amaranthus spp.		seeds and leaves
Pokeweed	Phytolacca americana	Aug–Nov	young leaves
Broomweed	Sida spp.		
Smartweed (Knotweed)	Polygonum spp.		seeds and plant
Pickerel Weed	Pontederia lanceolata	Jan–Feb, Oct–Dec	

Plant	Scientific Name	Harvest Time	Edible Parts
Buttonbush	Cephalanthus occidentalis		
Saltmarsh Mallows	Kosteletzkya virginica		okra-like vegetable
Pepper	Capsicum spp.		vegetable
Wild Pepper	Peperomia spp.		vegetable
Panicoid grass	Panicum/Setaria spp.	Oct-Nov	seeds

(A bibliography on the subject of *edible botanicals* is included at the end of the book.)

Of interest is a description by Dr. Strobel of preparation of coontie found in the Southeast:

> The land on our right consisted of the same pine sandy barren as I have already described. The Indian arrow root called coontie is found here in great quantities. We landed, and collected several roots, which were very large, weighing several pounds. This is the Indian's principle bread stuff. It is met with in most of the pine barrens in this section of Florida, but it grows in such profusion in this neighborhood, that they come from considerable distance to procure it.... The following is the manner of preparing it. A sufficient number of roots being collected, they are peeled, washed, and grated in the same manner as potatoes, and thrown in large tubs of water. After remaining in soak for a certain length of time, the water is stirred and strained; by this process it is freed of feculent matter. The coarse portion thus separated, may be given to hogs, while the finer portion, which passes through the sieve, is allowed to settle. The farina, which is almost insoluble in cold water, subsides at the bottom. The water is drawn off, and the yellow portions which remain on the top are removed. The white arrow root, which from its specific gravity, is found at the bottom, is collected, and repeatedly treated with freshwater, until it becomes perfectly pure, and white, of a granular, glistening, crystalline appearance. I am inclined to think, that when thus prepared, it is very nearly, if not quite equal to the Bermuda arrow root, not only as starch, but also as an article of diet.

Paleoethnobotanist Dr. Lee Ann Newsom notes that

> Citrus, banana, plantain, sugarcane, watermelon, peach, and wild lettuce all are "Old World" in origin, and did not arrive in the Americas until after European contact. In the same way, tomato comes from Mexico, and there is no evidence for its presence in Florida until after the Spanish arrival in the Sixteenth Century.... You might add maize (*Zea mays*) to the list; we do not believe it was grown prehistorically in the Florida peninsula (though it was cultivated in the panhandle region and possibly

as far south as just north of Tampa Bay…), but historic Seminole people did rely on maize, even as far south as Dade County.

Lopez de Velasco wrote a description of a unique method used in the Tequesta manatee hunt:

> In winter all the Indians go out to sea in their canoes, to hunt for sea cows. One of their number carries three stakes fastened to his girdle and a rope on his arm. When he discovers a sea cow he throws the rope around its neck, and as the animal sinks under the water, the Indian drives a stake through one of its nostrils, and no matter how much it may dive, the Indian never loses it, because he goes on its back. After it has been killed they cut open its head and take out two large [otic] bones, which they place in the coffin, with the bodies of their dead and worship them.

Robertson (1777) wrote an interesting paragraph regarding the Indians' physical attributes:

> None of them are deformed, or mutilated, or defective in any of their senses. All travelers have been struck with this circumstance, and have celebrated the uniform symmetry and perfection of their external figure…. The distresses and hardships of the savage life, which are often such as can hardly be supported by persons in full vigor, must be fatal to those of more tender age. Afraid of undertaking a task so laborious, and of such long duration, the women in some parts of America, extinguish the first sparks of that life which they are unable to cherish, and by use of certain herbs procure frequent abortions…. In such rude nations, such persons [as those young suffering defects or deformities] are either cut off as soon as they are born, or becoming a burden to themselves and to the community, cannot long protract their lives.

Of course, Robertson's statements are generalizations regarding all of America's indigenous population. However, since the southernmost part of the Floridan Peninsula was considered to be an extremely difficult environment in which to live, it would seem to be apropos. As Brother Villareal noted, "I and the others have constantly remained healthy, glory be to God, which helps us endure with little difficulty some of the burdens of the land that otherwise would seem insufferable."

And yet Robin Brown, writing in 1994 of the ancient skeleton of a 14 or 15 year old boy who had been born with spina bifida, noted that in spite of the terribly crippling illness, his people had "cared for him-they nursed him through the loss of a foot and carried him on their seasonal migrations. These ancient people were far from unfeeling savages whose actions were ruled only by expediency. Charity and kindness were here, 7,000 years ago."

During this first century and a half of European contact, both the French and English, as well as the Spanish, entered into the conquest of the New World. The former, Protestant Huguenots, attempted to establish Fort Caroline in the northern part of La Florida in an area that today is Port Royal, South Carolina. They attempted to expel Spain from Florida by attacking their headquarters at St. Augustine. A storm scattered the attacking fleet along the coast near today's Cape Canaveral. Don Pedro Menendez de Aviles took advantage of the situation, marched overland, captured the small Fort Caroline, and hanged the remaining men of the garrison over the age of 15. He then returned to St. Augustine, where on September 28th he located 208 of the starving stranded Frenchmen. Eight who claimed to be Catholic were spared. The remainder, all 200, were massacred. On October 10th, another group of 320 was found in the same area. Of these, some 150 surrendered. About 16 were spared for various reasons; the balance were massacred. The small coastal inlet where both incidents occurred became known as Matanzas, Spanish for slaughter.

In 1567 a small force sent from France recaptured Fort Caroline and retaliated by hanging the Spanish men of the garrison. The French then yielded their brief foothold in Florida for greener pastures elsewhere.

The English were establishing themselves in settlements along the more northern stretches of the Atlantic coast. They rapidly allied themselves with the indigenous people north of Florida. The English, too, coveted the peninsula and the people therein, especially since many of their Negro slaves were escaping into Spanish Florida, and there found some degree of refuge.

Disease epidemics took tremendous toll of the natives, both in the Caribbean Islands, the Keys, and on the mainland. By about 1600, yellow fever, typhus, measles, and smallpox had felled them by the thousands. Disease epidemics among the Native Americans have been recorded for the years 1612–1616, 1649, 1655, 1670, and 1672. One record states that by 1708, Ais and Tequestas numbered no more than several hundred. The Jeaga were so decimated that the few left were absorbed into the Calusa. Other remnant groups, now reduced to little more than bands, were the Mayaimi, Tancha, Muspa, Rio Seca, Santa Luce, Mayaca, and Jove (or Hobe).

Especially after the 1670 British founding of Charleston, raids by combined forces of Englishmen and members of the Creek Federation of Alabama and Georgia pressed farther and farther south. In the early 1700s, Creeks and Yuchis carried on warfare with the peninsular Indians as far south as Cape Florida (the tip of today's Key Biscayne, Miami–Dade

County). Their hunger for prisoners for slavery as well as new hunting grounds spurred them on. There they fraternized with pirates and ruthless slavers who would take every Indian they could lay hands on to sell to West Indian plantations, the hated English.

In 1675 the Bishop of Cuba, Gabriel Diaz Vara Calderon, visited Florida on a survey of missions. He noted that in South Florida there existed 13 tribes, one of which he called Vizcayanos, who more than likely were those formerly known as Tequestas. He commented that they lived at a large river (probably the Miami) at the Head of the Martires (the Keys) that flows into the Mayaimi Lagoon (Biscayne Bay). He reported that these savages were excellent swimmers and divers who salvaged everything useable from wrecked ships. Obviously, food and articles of clothing were highly desirable, as were metal knives, axes, and other durable materials, all of which helped make their lives more bearable in this difficult land. Perhaps the Indians had not realized the land was so inhospitable and unbearable until the Europeans arrived and offered them various goods to win their favor in return for friendship, trade, and conversion to Christianity (and slavery). Based on the archival data, corn was a premium item, and the degrees to which the natives would go to obtain it seemed to indicate diminution of their ancestral independence to fish, hunt, and forage. Gold, silver, copper, and iron began to make their way into the trade routes, all salvaged from or traded with the Europeans. None of the metals are found in any natural form on the lower peninsula. Swanton (1922) comments that:

> The culture along the Florida east coast generally seems to have belonged to that of Calos [Casique of the Calusas]. Its simplicity was partly due, without doubt, to the poverty of the country; in fact, in later times the economic condition was considerably advanced by frequent wrecks along the coast, though at the same time native industry must have been proportionately discouraged.

Jonathan Dickinson's journal of 1699 is a remarkable document that describes aboriginal life as he found it in the area north of Tequesta territory from Jupiter Inlet north to St. Augustine. It is known that his ship, the Reformation, was wrecked in 1696 about five miles south of Jupiter Inlet by the latitude worked out by its mate. Dickinson recorded the name of the Jeaga Indians town there as "Hoe-bay," derived from "Jeaga" which the Spanish called "Ho-ve." The English spelled it phonetically as "Hobe"; hence, today's name of the water body known as Hobe Sound. The Spanish "Hove" or "Jove" also meant Jupiter to the English, so the Inlet's name of Jupiter remains as well. Dickinson, his wife and baby, and 20 others were spared by the Hobe Indians whom they first encountered, and their travels took them among the Hobe (Xega or Jeaga), St. Lucia, and Jece (Ays or Ais). His descriptions of these people and their lifestyle more than likely closely paralleled life as it existed at that time among the Tequestas to the south. For example, he described that an important part of the formalities of religious ceremonies and other major occasions was preparation and consumption of the beverage known as the black drink. Boiling the leaves of caseena, the Yaupon holly (*Ilex vomitoria*), produced this strong emetic.

> Night being come and the moon being up, an Indian who performed their ceremonies, stood out, looking full at the moon, making a hideous noise, and crying out, acting like a madman for the space of half an hour, all the Indians being silent til he had done, after which they made a fearful noise, some like the barking of a dog, wolf, or other strange sounds; after this one got a log and set himself down holding the log upright on the ground and several others got about him, making hideous singing.... At length their women joined the concert ... which they continued til midnight.
>
> Dancing played an important part in many of their religious ceremonies. After ceremonial painting of their body, the Tequestans put on their belts and quivers and arrows and waited until the medicine men inaugurated the dance by shaking rattles and going through ritual procedure, then they began a stomping sort of dance which continued several hours, to the point of near exhaustion. Then they retired to the hut of the casique for the drinking of caseena. This continued for three days.

Romans describes preparation of caseena:

> The caffeine is by them used as a drink, they barbecue or toast the leaves and make a strong decoction of them; the men only are permitted to drink this liquor to which they attribute many virtues, and it is made so strong as to be black and raise a froth; when they drink it at their assemblies in the square, they call it black drink.

In 1702 Carolinian Thomas Nairne and his Indian allies destroyed both the fort and town of St. Augustine. He followed this two years later with attacks on Spanish missions across the northern half of the peninsula. His force's captives were sold into slavery in Charleston. Nairne's 1709 report to the Earle of Sunderland states:

> Your Lordship may perceive by the map that the garrison of St. Augustine is by this war Reduced to the bare walls their Castle and Indian towns all consumed Either by us in our Invasion of that place or by our Indian subjects, since who in quest of Booty are now obliged to go down as far

as the point of Florida as the firm land will permit they have drove the Floridians to the islands of the Cape, have brought in and sold many Hundreds of them, and Dayly now continue that trade so that in some years thay'le Reduce these Barbarians to a farr less number.

There remains not now, so much as one Village with ten houses in it in all Florida, that is subject to the Spaniards; nor have they any houses or cattle left, but such as they can protect by the Guns of their Castle of St. Augustine, that alone being now in their hands and which is continually infested by the perpetual Incursions of the Indians, subject to this Province.

In 1715, at the start of what would be known as the Yamassee War, Nairne, Indian agent for South Carolina, colonists, and many traders were killed at the hands of the vengeful Yamasees and their Apalachee, Savannah, and Lower Creek allies after years of mistreatment, indebtedness to a point of no possible return, enslavement, and endless conflict. The end of the two year, bloody war found the remnants of the Yamasees at the gates of Spanish St. Augustine.

Spanish Florida never would recover from these blows. With Spain's presence limited to the remains of St. Augustine, the Yamassee of South Carolina continued to decimate northern tribes of Florida, and made long raiding expeditions throughout the peninsula. Every Indian and escaped Negro slave they could capture was taken along the great slave trail that ran down the St. Johns and Kissimmee Rivers to Lake Okeechobee. They moved on down the Caloosahatchee to the Gulf Coast where they enslaved many Calusas. They raided down the east coast, destroying the last of the Jeaga and Ais, and on to Biscayne Bay to take their toll among the Tequestas. Here they met other slavers and pirates from the Bahamas with whom they caroused, all the while taking high numbers of Indians from the Everglades area into bondage. They and their English allies became hated enemies of the Florida tribes, who retreated ever more deeply into the southern wetlands interior of the peninsula. This strengthened the alliance between the remaining Spaniards and the natives to the point where, eventually, the latter became known as "Spanish Indians." Any Englishman unfortunate enough to fall into their hands most often was immediately put to death. They were the lucky ones; others were lashed to a wooden frame over a fire and burned to death.

In *The Southeastern Indians*, Charles Hudson notes that the slavery business was justified by a Carolinian, who wrote:

> It serves to lessen their number before the French can arm them, and it is a more Effectiall way of Civilizing and Instructing than all of the other efforts used by the French missionaries.
> It saved them from cruel deaths at the hands of their enemies.

Not all agreed. Another writer stated with obvious contempt:

> Through Covetousness of your gunns Powder and Shott and other European Comodities ... to ravish the wife from the Husband, kill the Father to get the Child and to burne and Destroy the habitations of these poor people into whose Country wee are Carefully received by them, cherished and supplyed when wee are weake, or at least have never done us hurt; and after wee have set them on worke to doe all these horrid wicked things to get slaves to sell the dealers in Indians call it humanity to buy them and keep them from being murdered.

A lengthy letter from the Governor of Cuba to the King of Spain laments:

> These tiny little tribes fight constantly and they are shrinking, as is testified to by the remembrance of the greater number there was twenty years ago, so that if they are left to their barbarous ways, they will have disappeared in a few years, whether on account of the skirmishes, or on account of the rum that they will drink until they burst, or on account of the children that they kill, or on account of those whom the small pox carries off for the lack of a remedy, and, finally, whether on account of those who perish at the hands of the Uchises.

As early as 1704 the Indians were moving south into the Keys, many all the way down to Key West, some even making their way to Havana, Cuba to escape their enemies. In 1711 the Bishop of Cuba managed to finance an expedition to rescue the Indians who had made their way to the southernmost tip of La Florida. Upon reaching the Keys, the expedition found the Indians engaged in a savage battle with the Yamasee. Over 2,000 of the desperate people had arrived to be saved, but the expedition's two small ships only could hold 270. This comparatively small group may have been the more unfortunate of all. In a short time after their arrival in Havana, more than 200 of the rescued people died of illness, and a very few, less than 20, later returned from whence they had come. Of those left behind in Florida, some more than likely were killed, others may have been pressed into bondage, some may have escaped via canoe to areas of the Gulf Coast, to Mesoamerica, or other islands of the Caribbean. Some apparently melted into the vastness of the Glades. Juan Elixio de la Puente (mistakenly) wrote "the Keys are without any of their former natives."

Archival records indicate that a party of Jesuits landed on Upper Matecumbe Key in 1743 in a last attempt to Christianize remnants of Tequesta, Calusa, Ais and any other tribes. The Indians stated they would renounce their customs, including polygamy and sacrifice of first-born children in exchange for food and alcohol from the Spaniards. The Jesuits built a mission, but a smallpox epidemic took a heavy toll on the Indians. Many survivors drank themselves to death. In the face of so hopeless a situation, within a few years the Jesuits abandoned their mission. Spanish troops razed the structure to keep it out of the hands of others, mainly pirates or the British, who might use it against them.

In 1755 Great Britain and France engaged in what was known as the French and Indian War. After it had continued for some six years, Spain was induced by France to enter the war; an unfortunate decision by the Spanish since it precipitated a British attack on and capture of Havana, Spain's capital of Cuba.

The three governments finally signed a peace treaty in 1763 in which England took Canada from France. Spain received the Louisiana Territory, and ceded all of Florida to England in exchange for the return of Havana. All Spanish residents of Florida were offered the opportunity to evacuate the peninsula, and the following year many did just that, joined by many south Florida Indians desperate to escape the hated English. They fared little better in their new land than did their fellow tribesmen who had preceded them.

The Gulf Coast Calusas were no more fortunate than any of the other indigenous people of the peninsula. As late as 1698 Calusas still were influencing neighbors, but within 150 years their culture was erased. Bernard Romans notes that the last of the Calusas, too, crossed to Havana in 1763. And yet, in spite of these comments another letter of 1763 states:

> The Indians of Ratones [Keys] and the south part of Florida cure great quantities of this fish [meros and Pardos] which, with hats and mats they make of grass and barks of trees in great perfection, they exchange in traffic with the Spanish who come from the Havana with European goods for the use of the natives.

Apparently, some of the surviving Tequestas, Calusas, and members of other small bands managed to continue to live in or near the Biscayne Bay area. There is evidence that fishermen from Havana were transporting them in their seasonal cycle from the Bay area to Cape Sable in September, thence to Key Vaca in the Keys, and later returning to the mouth of the Miami River. The travels were undertaken for the same reason they earlier had moved to Key Biscayne or the Honey Hill site: to take advantage of the availability of edible botanicals, fishing, and hunting.

Total conquest of La Florida was not quite complete; obviously some of the indigenous people managed to elude captivity, survived, traded, and fought enslavement and/or extermination in their own ways as demanded by the moment.

To the north, tribes of the Creek Federation continued their migration into the peninsula. Romans describes them as

> A mixture of the remains of Cawittas, Talapoosas, Coosas, Apalachias, Conshacs or Coosades, Oakmulgis, Oconis, Alibamons, Natchez, Weetumkus, Pakanas, Taensas, Chacsihoomas, Abekas, and some other tribes whose names I do not recollect.... They call themselves Muscokees and are at present known to us by the general name of Creeks, and divided into upper and lower Creeks....
>
> They are the next most numerous nation after the Choctaws.... This confederacy of remnants is a race of very cunning fellows, and with regard to us, the most to be dreaded of any nation on the continent, as well as their indefatigable thirst for blood [which makes them travel incredibly for a scalp or prisoner] as for their being truly politicians bred, and so very jealous of their lands, that they will not part with any, but endeavour constantly to enlarge their territories by conquest and claiming large tracts from the Cherokees and Chactaws.

Upper Creeks, who had lived along the Coosa and Tallapoosa branches of the Alabama River and spoke Muskogean, first settled in the more northern area. Today, that language that sometimes is referred to as 'Creek' is spoken mostly by the Seminoles of Brighton Reservation and Ft. Pierce. Those known as Lower Creeks, who had lived in the valleys of the Chattahoochee and Flint Rivers along the lower Alabama and Georgia borders and spoke Hitchiti, pressed further south, enjoying good relations with the English who were more generous with their gifts than the Spanish. Armed with muskets and other weapons supplied by the English, they raided their bitter Yamassee enemies who lived near Spanish settlements. The Yamassees eventually were exterminated except for those survivors who were enslaved and absorbed into their captors' bands.

The tribal origin of Lower Creeks is thought to be Oconee, whose home about the end of the 17th century was on the Oconee River in Georgia. They later moved to the Lower Creek country, then on south in the mid–18th century to the Alachua prairie in Florida, where they were joined by Sawokli, Tamathli, Appalachicola, Hitchiti, and Chiaha. Today the language of most Seminoles and Miccosukees is a dialect of the now extinct Hitchiti referred to as Mikasuki. Linguist and archaeologist Julian Granberry states that "all of Florida's coastal and lower central Indians originally came out of the Muskogean linguistic stock of present-day Alabama and Georgia, dating back as early as Paleo-Indian times. In historical times this included the Pensacola, Apalachees, Tocobaga, Calusa and Mayaimis, and on the lower east coast, the Tequestas, Jeagas, and Ais."

By the early 19th century, small parties of Lower Creeks hunted throughout the entire Florida Peninsula and made contact with Cuban fishermen with whom they traded for coffee, sugar, cigars, and rum.

Their trading was not confined to Florida. William Bartram (1774) describes the large canoes in which the Indians traveled

> quite to the point of Florida and sometimes across the gulph, extending their navigation to the Bahama islands and even to Cuba: a crew of these adventurers had just arrived, with a cargo of spiritous liquors, coffee, sugar, and tobacco.

However, the trade for white mens' goods, the metal materials of all kinds, beads, dyes, blankets, and guns with their flints and powder, reduced them to a state of dependency from which they were unable to return. As the Indians became more and more economically dependent on traders, they became more malleable in many ways.

While charting coastal waters in 1783, Spaniard Joseph Antonio de Evia reported talking with Yuchis, Tallapoosas, and Choctaws at Tampa Bay who were hunting deer for hides they hoped to trade with the English probably back farther north — sometimes as far as Georgia. Occasionally, on return, they attacked other villages and carried back anything they deemed of value as well as other Indians and negro slaves. Of course, this resulted in white retaliatory raids that forced the Indians even farther south.

Within ten years another Spaniard reported finding two seemingly semi-permanent villages in the same bay area. By 1822 only five villages were reported from Tampa Bay south.

Members of the many bands, including disgruntled or rogue whites and captured or runaway negro slaves, became known as Seminoles. The slaves, either male and female, were permitted to marry and while they remained in bondage, their offspring were born into the tribe as full members. Fairbanks writes

> Beginning about 1715, the wildest, most intransigent indians of the Southeast moved into Florida to become ... Seminole. In the Seminole Wars, it was again the wilder ... element that remained in Florida. By 1771, they were beginning to be referred to by a distinctive term, Seminole.

Bartram reports

> This handful of people possess vast territory, all East Florida and the greatest part of West Florida, which being naturally cut and divided into

thousands of islets, knolls, and eminences, by the innumerable rivers, lakes, swamps, vast savannas and ponds, form so many secure retreats and temporary dwelling places, that effectually guard them from any sudden invasions of attacks from their enemies and being such a swampy, hammocky country, furnishes such a plenty and variety of supplies for the nourishment of varieties of animals, that I can venture to assert, that no part of the globe so abounds with wild game or creatures fit for the food of man.

Thus they enjoy a superabundance of the necessaries and conveniences of life, with the security of person and property, the two great concerns of mankind. The hides of deer, bears, tigers and wolves, together with honey, wax and other productions of the country, purchase their cloathing, equipage and domestic utensils from the whites. They seem to be free from want or desires. No cruel enemy to dread, nothing to give them disquietude, but the gradual encroachments of the white people. Thus contented and undisturbed, they appear as blithe and free as birds of the air, and like them as volatile and active, tuneful and vociferous. The visage, action, and deportment of the Siminoles, form the most striking picture of happiness in this life.

Bartram's warning regarding the "gradual encroachments of the white people" in less than a quarter century would prove all too true.

In due course, remaining members of the old tribes lived independently in small groups or blended with the new, and collectively became known as Seminole/Mikasukis.

In 1776 America began its war of independence; British rule over Florida came to an end with that war's end and the 1783 Treaty of Paris. The 1790 Continental Congress attempted to bring a degree of fairness to the indigenous people by passing *An Act to regulate trade and intercourse with the Indian Tribes* (see Appendix I.) to which few seemed to have paid attention. Spain again took possession of La Florida as a result of the treaty, only to decide to sell the peninsula, comprised of the provinces of East and West Florida, to the United States. It was ceded to the United States under provisions of the *Treaty of Amity, Settlement, and Limits, between the United States of America and His Catholic Majesty*, of February 22, 1819. Acceptance on the part of the United States occurred on February 19, 1821. Some 16 months later Andrew Jackson formally took possession of the territory for the United States.

Two years prior to that, the First Seminole War (1816–1818) had begun as a result of General Andrew Jackson's raids on blacks affiliated with the Indians and on Indian villages along the north Florida border in reprisal for Indian raids upon settlers. Those hostilities terminated with the United States taking possession of that land.

The Second Seminole War (1835–1842) was a result of the Treaty of Moultrie Creek (see Appendix II., Treaty with the Florida Tribes of Indians). The 1823 Treaty of Moultrie Creek established the Government's policy of removal of Indians to Oklahoma reservations. Here an Indian who was not a traditional Creek leader, but a warrior whose natural abilities placed him in that role: Asiyaholi, or Osceola. Jerald Milanich, in his article about him in *Archaeology* magazine, relates:

> In 1835, General Wiley Thompson, the government agent to the Seminole Indians, politically humiliated Osceola by placing him in chains after he refused to sign a treaty agreeing to leave Florida. But Osceola would have his revenge. On the afternoon of December 28, 1835, the Seminole chief and some fifty of his warriors ambushed Thompson outside Fort King, a U.S. Army outpost in central Florida. Thompson was shot fourteen times and then scalped; an army lieutenant and three settlers were also killed. Osceola's ambush of Thompson touched off the Second Seminole War, one of the bloodiest conflicts ever fought in the long history of Indian wars.

The army forced relocation of some 3,000 Seminoles to a territory west of the Mississippi (today's Oklahoma). With them were 500 blacks, other blacks known as Maroons who were born free in Florida and were descendants of runaway slaves, and slaves. The Seminoles were joined by Creek, Choctaw, Chickasaw, and Cherokee (referred to as the five civilized tribes) in the move that resulted in terrible loss of property, suffering and lives. The road to the new country in Cherokee was called *Nunna daul Tsunyi*, which translates to "the trail where we cried," better known today as "the trail of tears." The Second Seminole War ended without formal declaration or treaty between the Seminoles and the U.S. Government. After eight years, an estimated 300 holdouts had disappeared deep into the Florida swamps.

The Third Seminole War (1855–1858) followed in ten years, (see Appendix III., Treaty With The Seminole) stimulated by destruction of an Indian garden on a hammock in the Big Cypress. The army refused a request for reparations, so Miccosukee Seminole leader Billy Bowlegs and a band of his warriors attacked army surveyors in the Big Cypress in retaliation. The army tracked down and sent west almost half of the remaining Indians, then gave up on the rest hidden in the vastness of the Everglades domain. This Third Seminole War, too, ended without formal treaty.

Florida became a state within the Union in 1845, and continued the process of removal of the Indians to western territories. Three Indian wars later, each very costly in dollars and lives on both sides, many Seminoles had been forced to the Oklahoma Territory. However, like the earlier Calusas, Tequestas, and others before them, again some remained; proud, determined, and defiant, they would take no more. Perhaps their position was reinforced by the groups of "Spanish Indians" to whom they referred at that time as still living deep within the Glades. Former State Archaeologist Vernon Lamme notes that, back in the early 1930s, he first heard of a group of Indians in the Glades who were not Seminoles. Ten years later "Josie Jumper, a Miccosukee from near Turner River," who then lived with his family on the Dania Reservation, "told me of an Indian, smaller than Seminoles, who had thrown a spear with a crude flint projectile point, striking his mother in the thigh. He says he still has the spearhead." Lamme speculates that that Indian "might possibly be the last of the Caloosa tribe which inhabited Florida at the time of Ponce de Leon, DeSoto, Narvaez and other intrepid explorers."

In spite of the armies and funding of an entire nation, some of the indigenous people were able to elude resettlement. It is entirely possible that during the 18th and 19th centuries those few remnants that remained of South Florida's earliest named tribes or bands intermarried with people of Sawokli, Tamathli, Hitchiti, Chiaha, Yuchi, Yamassee, Apalachee, Timucua, and Creek descent, and any others who yearned for space free from the expanding white population. This included escaped and Indian enslaved negroes, and in spite of being forbidden by the tribe, some whites. The Indians brought with them their gods, ceremonies and culture. For example, in Florida when corn ripens in early June after winter planting, the people gather from scattered camps to associate with old friends, renew acquaintances and clan ties, and share gossip. Dances and socializing precede the season of *Buskita*, derived from the Creek word *poskita* meaning to fast, fasting being one of the principal means of attaining purity among Southeastern Indians. The purpose of the *poskita* was to ensure the bounty of nature as well as the Creek social order in the eyes of their (and Seminole) all-powerful god Breathmaker. According to Sturtevant, this celebration marks the beginning of the new year that is based on an annual cycle, the date of which is determined by the medicine man. Rituals of fasting and self-purification provide for the general health and well-being of the people. The secret sacred medicine bundle is re-examined to ensure its potency and power for good. A new fire is kindled with four logs set in the cardinal directions as a symbol of the tribe. Lighting of the fire symbolizes renewal of the Seminole circle of life. They view the whole universe spinning slowly in a circle like the logs in the ceremonial fire. What was, will be and will cease to be again. The celebration also provides an annual court for consideration of violation of tribal customs. Important tribal issues are discussed. In general, it is a time of renewal of matrilineal clan ties and tribal identity. This significant expression of culture is both a sacred and secular activity that is very much alive to this day. Collectively these hardy people became known as Seminole/Mikasukis. They have remained in their watery wilderness domain to the present day as two federally recognized tribes, the Seminole Tribe of Florida in 1957, and the Miccosukee Tribe of Indians in 1962. An in-depth study of these two tribes of Native American people of the southeastern United States is found in *Unconquered People: Florida's Seminole and Miccosukee Indians* by Brent Richards Weisman (1999). The book is one of a series published by University Press of Florida *of Native Peoples, Cultures, and Places of the Southeastern United States*, edited by Jerald T. Milanich.

Postscript

Having researched the foregoing for the past 15 years, I trust I am finally able to provide a response to my fellow Rotarians regarding many facets of the prehistory of South Florida. Obviously man's time here predated by 13 millennia or more the 1800s and the wealthy, energetic, and capable Henry Flagler and his incredible railroad.

The archaeological record reveals that dinosaurs never roamed the peninsula we know today as Florida. Their time on earth was from about 200 million to 80 million years ago, when Florida probably was nothing more than the floor of an ancient sea. According to the fossil record, 78 million or more years passed before man's earliest ancestors were to leave their bones to fossilize in the dust of the African continent.

Another 77,985,000 years were to pass before modern man, as Paleoindians, followed the hunt and his botanicals northward out of Asia and Europe to Siberia and, eventually, across the temporarily exposed Bering land bridge to the North American continent. They were the only humans, those Paleoindians, to be found on the North and South American continents. The archaeological record offers no indication whatsoever that earlier hominids evolved on or had made their way to these continents independently.

However, recent research and data based on improved dating techniques does suggest that people may have ventured onto the new continent as early as 30,000 years ago, possibly by boat across the Pacific to the west coast, and/or by boat from Europe's Iberian Peninsula, along coastal ice, across the Atlantic to the east coast, both fairly recent hypotheses that await confirmation.

The Floridan Peninsula was about as far east and south on the North American continent as one might wander through a mostly frigid land fraught with what to us today would be unimaginable dangers at every turn. There were many smaller creatures, easy game, to be hunted for sustenance. But there were many larger ones too, some with similar ideas in mind, with an eye and taste for this human fair game. Evolution had placed above man in the food chain such fearsome end-of-the-Ice-Age beasts as dire wolves, saber-toothed tigers, and the huge American lion, to name but a few. And so, understandably, man's wanderings across the icy continent from whatever directions took many, many generations over thousands of years.

By about 12,000 years ago, Paleoindians ventured onto the Floridan Peninsula. Most of the very large species of Ice Age animals of that time had become, or were becoming extinct, but enough still were around to make daily living hazardous at best.

The sea was much lower, with more of the Floridan landmass exposed above its surface than today. The climate was drier, so the land more closely resembled modern African savannas. On the southeast coastal area, man sought refuge from the elements along the bulwark of the low, stony Atlantic Ridge, and in some of the solution holes that dot the area. He hunted and defended himself with rocks, throwing sticks, clubs, spears, lances, and atlatls. Suitable stone materials such as flint and quartz were not available in this new land. Chert and agatized coral provided adequate substitutes in the central west coast area. Elsewhere, wood and other plant materials, bone, animal parts, and shell became the mainstays of his tool-making and weapons industries. His very survival was predicated upon his ability to adapt to the slowly changing, rather unique environment.

Changing weather and environmental patterns did occur on grand scale as the ice to the north melted and the seas flooded over coastal areas of the low, flat land of the Floridan Peninsula. The climate became wetter and warmer, the peninsula began to more closely resemble the sub-tropical environment we know today. This certainly did not happen all at once. Rather, the sea level and weather fluctuated a number of times over the millennia, and people migrated north or south in response to the land mass fluctuations that, in turn, altered availability of animals and botanicals to hunt for food.

Obviously, botanicals fluctuated as well, but the archaeological record of many of the plants and their use potential is incomplete. New methods being devised by archaeobotanists daily are adding interesting data to this facet of ancient life.

As the environment stabilized, populations of ancient man tended to become more stable as well. Both the people and the animals enjoyed a more ready supply of freshwater and food. Consequently, the need for lengthy migrations minimized, and both were able to reduce their wanderings to that dictated by seasonality of foodstuffs.

The more sedentary life apparently agreed with the early native Floridians. Their numbers increased. The advent of the canoe provided increased mobility through this watery glades realm; the people even ventured forth on the sea and became skilled at food gathering therein. They became adept at the making, tempering, and firing of pottery almost 4,000 years ago, an art that had taken some eight millennia to reach them, and would take another two millennia to travel to other indigenous people farther north. Conversely, it would take roughly an equal several thousand years for the technology of the bow and arrow to travel from northern woodlands down through the peninsula. Useful material objects such as stone, tools, points, and pottery began to be traded. Those are the durable items that survive most readily in the archaeological record. However, Michael Cook in his *Brief History of the Human Race* admonishes: "Focusing on a culture's most durable artifacts may unwittingly distort that culture's values." Certainly hides and other perishables entered into barter as well. Twisted fiber cord and products of the early weavers' art entered the picture too. While organic materials generally disintegrate rapidly, enough have been preserved deep within anaerobic muck, and as impressions on ceramics, to provide a picture of their manufacture and use.

Before touching briefly on European incursion and its effects on the people of Florida, it is important to recognize the extent of the preceding discussions involving recovery of artifacts of material culture from the peninsula's wetlands. The Florida Anthropological Society states: "There are very few places in the world where a 10,000 year continuous record of cultural and environmental data is preserved." This unique record of human history has been preserved in Florida's wetlands, entombed in anaerobic permanently waterlogged deposits. The Society states:

> The heritage of Florida's wetlands provides extensive information about [1] flora, fauna, and climate, [2] human utilization of resources for food, fiber, and artifacts, [3] human skeletons permitting studies of age at death, injuries, nutrition, pathologies, and DNA and MtDNA evaluations of preserved human brain tissue, [4] artistic expressions (wooden totems, masks, figurines; wood and bone carvings), and [5] more than 200 Indian canoes. These seldom preserved organic remains make up at least 90% of all cultural inventories.

As man's numbers increased and they coalesced into bands, groups, and chiefdoms, it became necessary to enter into some form of government in order to accomplish major activities. Five centuries ago, when Europeans first set foot upon this land they called La Florida, some groups of Native Americans were governed by a chief, or casique, the latter a term adopted by the Spaniards from a native language in Hispaniola. It would seem apparent that this political system had been in place for a very long time, and these chiefdoms had enabled projects of incredible proportion to have been systematically accomplished. The people had successfully raised themselves above threatening waters, or water that provided security, by building great mounds with materials at hand.

Canals had been dug to facilitate canoe travel from one place to another. The work of artisans in local materials flourished. Improved nutrition and alteration of lifestyle had increased the people's stature, possibly inches in height, and they developed a more robust physical carriage than that of their forebears. Life spans increased from the 20 or 30 years of the early people to 40 and 50. They seemed to have been doing pretty well for their day and place.

One can only visualize the physical appearance of Florida's earliest people. The sole collection of existing portraits are from the work of Jacques Le Moyne, an artist commissioned to record the history of the French colony established in 1564 on the St. Johns River. The colony was destroyed by the Spanish after only a year, but LeMoyne survived to years later create from memory 42 drawings that document mid-16th century Indian life. They are considered flawed due to LeMoyne's insistence on creating idealized human forms. In 1591 Theodore deBry prepared them as engravings for publication as the *Narrative of LeMoyne surnamed DeMorgues*. They are a remarkable record of Timucuan life in the northern part of Florida. However, they must not be construed to represent all people in the peninsula. Many Europeans of that time were rather small in stature (Ponce stood four feet, nine inches tall) and when they first saw these Native American Indians they recorded many of them as being giants. Small wonder the aborigines stood tall in the eyes of the Spaniards.

Tall or no, it took a comparatively few years for the better armed Europeans, who carried diseases to which the aborigines had no immunity, to destroy almost everything human that had evolved on the peninsula for some 12,000 years. Other Native Americans quickly migrated south onto the peninsula to fill the void. Even with their incredible ecological adaptation, there was no way they, anymore than the original inhabitants, could withstand the overwhelming organization, superior weapons, diseases, and continually growing numbers of Europeans.

European nations fought against each other, ostensibly for God and King in this far-from-home new land, for what to them was rich and fertile new territory. As the polity of the land evolved from colonies in territories to a new nation, the Indian migrants continued to be eliminated or relocated to make way for the swelling population of a new country. A plea from Seminole leader Eneah Emathla fell on deaf ears:

> We hope that you will not send us south to a country whither neither the hickory nut, the acorn, nor the persimmons grow; we depend much upon these productions of the forest for food; in the south they do not grow.

At least most of them were relocated. According to the Seminole/Miccosukees of that time, remnants from the earlier indigenous groups were said to still be hidden deep within the wilderness called Everglades. About 1904 botanist John Gifford visited a settlement of unfriendly Indians in the southern Everglades (within today's Everglades National Park). He saw pumpkins hanging from live oaks, while under the trees were bananas and a kind of dasheen, taro-like plants, probably coontie. Gifford speculated that they might have been a Calusa remnant in spite of the belief that last members of that tribe had been absorbed by the mid–19th century as part of the Seminoles. Apparently some of them were able to join the people of the Upper and Lower Creek tribes that had migrated south, possibly other tribe remnant members that had moved north from the Keys, and live in south Florida today as two separate, healthy growing tribes, the Seminoles and Miccosukees.

Finally, along came Henry and his railroad.

Glossary

amulet— anything hung around the neck, wrists, arms or legs as an imagined preservation against illness, witchcraft or other evils.

anthropology— the science of the study of man in all his leading aspects— physical, mental, historical — to investigate the laws of his origin and progress, to ascertain his place in nature and his relation to inferior forms of life.

aquifer— water–bearing rock formation.

artifact— an object produced or shaped by human workmanship, especially a tool, weapon or ornament of archaeological or historical interest.

atlatl— an Aztec word for a spear thrower that effectively serves to lengthen the thrower's arm, thereby increasing the striking force of the projectile.

bannerstone— a sliding weight on the shaft of an atlatl designed to further increase the force of the propelled spear.

Beringia— the ancient land mass bared by Ice Age receding seas that connected Siberia and the Kamchatka Peninsula east to the Aleutian Islands and Alaska.

biota— animal and plant life of a particular region considered as a total ecological entity.

bipoint—formed to a point on both ends.

bryozoans— small aquatic animals of the Phylum Bryozoa that reproduce by budding and form moss-like or branching colonies.

bundle burial— bones of the dead which were disjointed, defleshed, cleaned, bundled, sometimes dyed red, often wrapped in a hide prior to burial.

caseena— an hallucinogenic "black drink" steeped from the parched leaves of the Yaupon Holly (*Ilex vomitoria*).

Casique— an Arawak language title for chief or leader transcribed by the Spaniards from the Taino Indians of Hispaniola.

celt— an axelike tool, generally defined as lacking perforations or grooving for hafting.

cenote— from the Maya "dzonot" meaning a well or sinkhole in the earth generally containing freshwater.

charnel— a chamber or building in which bodies or bones are deposited.

chert— a mineral variety of silica, usually formed as nodules in limestone. Heat-treated and flaked or knapped to form points, knives, scrapers and other tools.

columella— the central columnlike structure of a Gastropod shell that is removed to increase the internal volume of the shell for use as a dipper or cup. The columella itself was utilized as a plummet, hammer, chisel, awl, or other tool.

context— consists of the matrix (surrounding material), provenance (the horizontal and vertical position within the matrix), and association with other artifacts usually in the same matrix.

coprolite— fossilized animal or human waste.

cultural— pertaining to characteristic features of and behavior typical of a group or class; in this case, humans.

debitage— lithic wastes left from tool-making.

dolomite— a light colored gray, pink or white essentially calcium carbonate magnesia-rich sedimentary rock.

ethnology— the anthropological study of socio-economic systems, cultural heritage and traditions, especially of cultural origins and factors influencing primitive societies.

euryhaline— tolerant of fresh and saltwater.

extant— still in existence; not destroyed, lost or extinct.

extinct— no longer existing in living form.

feature— a non-portable artifact, such as architectural elements, hearths, or soil stains.

Glossary

flexed burial— the body or skeleton placed in its grave on its side in a bent or fetal position.

fossil— a remnant or trace of an organism of a past geological age such as a footprint, skeleton, shell or leaf imprint embedded in the earth's crust.

geology— the study of the origin, history and structure of the earth.

gorget— a decorative collar or armor protective of the throat.

graver— an engraver's cutting tool.

hammerstone— a heavy stone used as a hammer.

hydric— very moist or wet, as in wetlands.

karst— the porous limestone foundation of the Floridan Peninsula.

knap— to strike sharply. To break or chip with a sharp blow. For example, to chip or work chert, flint or other stone into points and tools.

labret— an ornament inserted into a perforation in the lip.

lithic— relating to or characteristic of a stage in man's use of stone as a cultural tool. Of, relating to, or made of stone in Greek.

material culture— the aggregate of a society's physical objects or artifacts.

mesic— moderately moist.

Mesoamerica— that geographic landmass extending from central Mexico south to the Pacific coast of El Salvador.

microlith— a small handtool or point fashioned of stone.

millennium— a time span of 1,000 years.

non-artifactual remains— organic and environmental materials, such as animal bones, plant remains, sediments, and soils.

ooid— a tiny calcareous grain or sphere formed around a foreign core.

oolite— limestone rock formed of ooids.

paleo— of or dealing with ancient forms.

paleobotany— the study of plant fossils and ancient vegetation.

paleolithic—from the Greek words "paleo" for ancient and "lithic" for stone; the cultural period beginning with the earliest chipped stone tools, about 1,500,000 to 15,000 years ago.

paleontology— the study of fossils and ancient life forms.

palynology— identification of plants through pollen analysis.

Pangaea— the ancient super landmass that separated into North America and Africa according to the theory of continental drift.

paradigm— an example, illustration or model.

Pleistocene— 2,000,000 to 10,000 years ago, the geological time commonly referred to as the Ice Age.

plummet— anything that weighs down or oppresses; in archaeological terms, a shaped bone, stone or other heavy material that generally was hung around the neck, arms or legs by a thong or cord.

prehistory— the history of mankind during the period prior to recorded events- to about A.D. 1500 in North America.

primary burial— the body is placed directly in a burial mound without first being stored in a charnel house.

profile— the vertical face of an archaeological site that reveals various layers and inclusions.

radiocarbon dating— dating of remains of organic origin. Living organisms absorb radioactive carbon as they grow. After death, the radioactive carbon decays at a known rate. The approximate time of death of the organism can be determined by measuring the proportion of remaining radioactive carbon.

shard— also sherd. A piece of broken pottery.

stalactite— a mineral deposit, usually of calcium or aragonite, projecting downward from the roof of a cavern as a result of dripping mineral rich water.

stalagmite— a mineral deposit growing upward from the floor of a cavern as a result of dripping mineral rich water.

steatite— a white to green talc, also called "soapstone," that is soft and workable into bowls, beads, etc.

stratigraphy— the arrangement of strata, horizontal layers of accumulated material each of which contains specific types of artifacts delineating a time frame within a chronological sequence.

tectonic— pertaining to or resulting from deformation of the earth's crust.

type— things sharing common traits or characteristics that distinguish them as an identifiable group or class. An example or model.

typology—the systematic organization of artifacts into types based upon shared attributes.

xeric—of, characterized by, or adapted to an extremely dry habitat. Low or deficient in available moisture for the support of life.

Appendix I

An Act to Regulate Trade and Intercourse with the Indian Tribes

UNITED STATES STATUTES AT LARGE 1:136-138, 1790

Section 1. *Be it enacted by the Senate and House of Representatives of the United States of America in Congress assembled,* That no person shall be permitted to carry on any trade or intercourse with the Indian tribes, without a license for that purpose under the hand and seal of the superintendent of the department, or of such other person as the President of the United States shall appoint for that purpose; which superintendent, or other person so appointed, shall, on application, issue such license to any proper person, who shall enter into bond with one or more sureties, approved of by the superintendent, or person issuing such license, or by the President of the United States, in the penal sum of one thousand dollars, payable to the President of the United States for the time being, for the use of the United States, conditioned for the true and faithful observance of such rules, regulations and restrictions, as now are, or hereafter shall be made for the government of trade and intercourse ,with the Indian tribes. The, said superintendents, and persons by them licensed as aforesaid, shall be governed in all things touching the said trade and intercourse, by such rules and regulations as the President shall prescribe. And no other person shall be permitted to carry on any trade or intercourse with the Indians without such license aforesaid. No license shall be granted for a longer term than two years, *provided nevertheless,* That the President may make such order respecting the tribes surrounded in their settlements by the citizens of the United States, as to secure an intercourse without license, if he may deem it proper.

Sec. 2. *And be it further enacted,* That the superintendent, or person issuing such license; shall have full power and authority to recall all such licenses as he may have issued, if the person so licensed shall transgress any of the regulations or restrictions provided for the government of trade and intercourse with the Indian tribes, and shall put in suit such bonds as he may have taken, immediately on the breach of any condition in said bond: *Provided always,* That if it shall appear on trial, that the person from whom such license shall have been recalled, has not offended against any of the provisions of this act, or the regulations prescribed for the trade and intercourse with the Indian tribes, he shall be entitled to receive a new license.

Sec. 3. *And be it further enacted,* That every person who shall attempt to trade with the Indian tribes, or be found in the Indian country with such merchandise in his possession as are usually vended to the Indians, without a license first had and obtained, as in this act prescribed, and being thereof convicted in any court proper to try the same, shall forfeit all the merchandise so offered for sale to the Indian tribes, or so found in the Indian country, which forfeiture shall be one half to the benefit of the person prosecuting, and the other half to the benefit of the United States.

Sec. 4. *And be it enacted and declared,* That no sale of lands made by any Indians, or any nation or tribe of Indians within the United States, shall be valid to any person or persons, or to any state whether having the right of pre-emption to such lands or not, unless the sale shall be made and duly executed at some public treaty, held under the; authority of the United States.

Sec. 5. *And be it further enacted,* that if any citizens or inhabitants of the United States, or either of the territorial districts of the United States, shall go into any town, settlement or territory belonging to any nation or tribe of Indians, and shall there commit any crime upon or trespass against, the person or property of any peaceable and friendly Indian or Indians, which, if committed within the jurisdiction of any state, or within the jurisdiction of either of the said districts, against a citizen or white inhabitant thereof, would be punishable by the laws of such state or district, such offender or offenders shall be subject to the same punishment, and shall be proceeded against in the same manner as if the offense had been committed within the jurisdiction of the state or district to which he or they may belong, against a citizen or white inhabitant thereof.

Sec. 6. *And be it further enacted,* That if for any of the crimes or offenses aforesaid, the like proceedings shall be had for the apprehending, imprisoning or bailing of the offender, as the case may be, and for recognizing the witnesses for their appearance to testify in the case, and where the offender shall be committed, or the witnesses shall be in a district other than that in which the offence is to be tried, for the removal of the offender and the witnesses or either of them, as the case may be, to the district in which the trial is to be had, as by the act to establish the judicial courts of the United States, are directed for any crimes or offences against the United States.

Sec. 7. *And be it further enacted,* That this act shall be in force for the term of two years, and from thence to the end of the next session of Congress, and no longer.

Approved, July 22, 1790

Note: Sec. 4 regarding sale of lands—traditionally, land is sacred, its source being the Breathmaker, a Seminole deity who entrusts it to the tribe or group to utilize and enjoy so long as it is not changed or altered in any way. Thought of individual ownership is completely foreign, but individuals have in common with their fellow tribesmen the right to live, forage, and hunt thereon, to exclusion of all unfriendly people.

(Author's note: Of particular interest in this 1790 non-intercourse act with Indian tribes is that it calls for no Indian land cessions without specific approval of the United States.)

Appendix II

Treaty with the Florida Tribes of Indians (Moultrie Creek)

UNITED STATES STATUTES AT LARGE 7:224-228, 1823.

ARTICLE 1. The undersigned chiefs and warriors, for themselves and their tribes, have appealed to the humanity, and thrown themselves on, and have promised to continue under, the protection of the United States, and of no other nation, power, or sovereign; and, in consideration of the promises and stipulations hereinafter made, do cede and relinquish all claim or title which they may have to the whole territory of Florida, with the exception of such district of country as shall herein be allotted to them.

ARTICLE II. The Florida tribes of Indians will hereafter be concentrated and confined to the following metes and boundaries: commencing five miles north of Okehumke, running in a direct line to a point five miles west of Setarky's settlement, on the waters of Amazura, (or Withlahuchie river.) leaving said settlement two miles south of the line; from thence, in a direct line, to the south end of the Big Hammock, to include Chickuchate; continuing, in the same direction, for five miles beyond the said Hammock — provided said point does not approach nearer than fifteen miles the sea coast of the Gulf of Mexico; if it does, the said line will terminate at that distance from the sea coast; thence, south, twelve miles; thence in a south 30% east direction, until the same shall strike within five miles of the main branch of Charlotte river; thence, in a due east direction, to within twenty miles of the Atlantic coast; thence, north, fifteen west, for fifty miles and from this last, to the beginning point.

ARTICLE III. The United States will take the Florida Indians under their care and patronage, and will afford them protection against all persons whatsoever; provided they conform to the laws of the United States, and refrain from making war, or foreign nation, without having first obtained the permission and consent of the United States: And, in consideration of the appeal and cession made in the first article of this treaty, by the aforesaid chiefs and warriors, the United States promise to distribute among the tribes, as soon as concentrated, under the direction of their agent, implements of husbandry, and stock of cattle and hogs, to the amount of six thousand dollars, and an annual sum of five thousand dollars a year, for twenty successive years, to be distributed as the President of the United States shall direct, through the Secretary of War, or his Superintendents; and Agent of Indian Affairs.

ARTICLE IV. The United States promise to guaranty to the said tribes the peaceable possession of the district of country herein assigned them, reserving the right of opening through it such roads, as may, from time to time, be deemed necessary; and to restrain and prevent all white persons from hunting, settling, or otherwise intruding upon it. But any citizen of the United States, being lawfully authorized for that purpose, shall be permitted to pass and repass through the said district, and to navigate the waters thereof, without any hindrance, toll, or exaction, from said tribes.

ARTICLE V. For the purpose of facilitating the removal of the said tribes to the district of country allotted them, and, as a compensation for the losses sustained, or the inconveniences to which they may be exposed by said removal, the United States will furnish them with rations of corn, meat, and salt, for twelve months, commencing on the first day of February next; and they further agree to compensate those individuals who have been compelled to abandon improvements on lands, not embraced within the limits allotted, to the amount of four thousand five hundred dollars, to be distributed among the sufferers, in a ratio to each, proportional to the value of the improvements abandoned. The United States further agree to furnish a sum, not exceeding two thousand dollars, to be expended by their agent, to facilitate the transportation of the different tribes to the point of concentration designated.

ARTICLE VI. An agent, sub-agent, and interpreter, shall be appointed, to reside within the Indian boundary aforesaid, to watch over the interests of said tribes; and the United States further stipulate, as an evidence of their humane policy towards said tribes, who have appealed to their liberality, to allow for the establishment of a school at the agency, one thousand dollars per year for twenty successive years; and one thousand dollars per year, for the same period, for the support of a gun and blacksmith, with the expenses incidental to his shop..

ARTICLE VII. The chiefs and warriors aforesaid, for themselves and tribes, stipulate to be active and vigilant in the preventing the retreating to, or passing through, of the district of country assigned them, of any absconding slaves, or fugitives from justice; and further agree, to use all necessary exertions to apprehend and deliver the same to the agent, who shall receive orders to compensate them agreeably to the trouble and expenses incurred.

ARTICLE VIII. A commissioner, or commissioners, with a surveyor, shall be appointed, by the President of the United States, to run and mark, (blazing fore and aft the trees) the line as defined in the second article of this treaty, who shall be attended by a chief or warrior, to be designated by a council of their own tribes, and who shall receive, while so employed, a daily compensation of three dollars.

ARTICLE IX. The undersigned chiefs and warriors, for themselves and tribes, having objected to their concentration within the limits described in the second article of this treaty, under the impression that the said limits did not contain a sufficient quantity of good land to subsist them, and for no other reason; it is therefore, expressly understood, between the United States and the aforesaid chiefs and warriors that, should the country embraced in said limits, upon examination by the Indian agent and the commissioner, or commissioners, to be appointed under the 8th article of this treaty, be by them considered insufficient for the support of the said Indian tribes; then the north line, as defined in the 2nd article of this treaty. Shall be removed so far north as to embrace a sufficient quantity of good tillable land.

ARTICLE X. The undersigned chiefs and warriors. for themselves and tribes, have expressed to the commissioners their unlimited confidence in their agent, Col. Gad Humphreys, and their interpreter, Stephen Richards, and, as an evidence of their gratitude for their services and humane treatment, and brotherly attentions to their wants, request that one mile square, embracing the improvements of Enehe Mathla, at Tallahassee (said improvement:! to be considered as the centre) be conveyed, in fee simple, as a present to Col. Gad Humphreys. And they further request, that one mile square, at the Ochesee Bluffs, embracing Stephen Richard's field on said Bluffs, be conveyed in fee simple, as a present to said Stephen Richards. The commissioners accord in sentiment with the undersigned chiefs and warriors, and recommend a compliance with their wishes to the President and Senate of the United States; but the disapproval, on the part of the said authorities, of this article, shall, in no wise, affect the other articles and stipulations concluded on in this treaty.

In testimony whereof, the commissioners, 'William P. Duval. James Gadsden, and Bernard Segvui, and the undersigned chiefs and warriors, have hereunto subscribed their names and affixed their seals, Done at camp on Moultrie creek, in the territory of Florida, this eighteenth day of September, one thousand eight hundred and twenty-three, and of the independence of the United States the forty-eighth.

William P. Duval, [L. s.]
James Gadsen, [L. s.]
Bernard Segui, [L. s.]
Nea Nathia, his x mark [L. s.]
Tokose Mathla, his x mark [L. s.]
Ninnee Homata Tustenuky, his x mark, [L. s.]
Miconope, his x mark [L. s.]
Nocosee Ahola, his mark [L. s.]
John Blunt, his mark [L. s.]
Otlcmata, his x mark, [L. s.]
Tuskeeneha, his x mark, [L. s.]
Tuski Hajo, his x mark, [L. s.]
Econchatimico, his x mark, [L. s.]
Emoteley, his x mark, [L. s.]
Mulatto King, his x mark, [L. s.]
Chocholohano, his x mark, [L. s.]
Ematlochee, his x mark, [L. s.]

Wokse Holata., his x mark, [L. s.]
Amathla Ho, his x mark, [L. s.]
Holatefiscico, his x mark, [L. s.]
Chefiscico Hajo, his x mark, [L. s.]
Lathloa Mathla, his x mark, [L. s.]
Senufky, his x mark, [L. s.]
Alak Hajo, his x mark, [L. s.]
Fahelustee Hajo, his x mark, [L. s.]
Octahamico, his x mark, [L. s.]
Tusteneck Hajo, hi x mark [L. s.]
Okoskee Amathla, his x mark, [L. s.]
Ocheeny Tustenukv, his x mark, [L. s.]
Phillip: his x mark, [L. s.]
Charley Amathla, his x mark [L. s.]
John Hoponey, his x mark, [L. s.]
Rat Head, his x mark, [L. s.]
Holatta Amathla, his :t mark, [L. s.]
Foshatchimico, his x mark, [L. s.]

Signed, sealed, and delivered, in the presence of —

George Murray, secretary to the commission C. D'Espinville, lieutenant, Fourth Artillery G. Humphreys, Indian agent, Jno. B. ScoU, lieutenant, Fourth Artillery Stephen Richards, interpreter, Isaac N. Cox William Travers, J. Erving, captain, Fourth Artillery, Horatio S. Dexter. Harvey Brown, lieutenant, Fourth Artillery

ADDITIONAL ARTICLE

Whereas Neo Mathla, John Blunt, Tuski Hajo, Mulatto King, Ematb lochee, and Econchatimico, six of the principal Chiefs of the Florida Indians, and parties to the treaty to which this article has been annexed, have warmly appealed to the Commissioners for permission to remain in the district of country now inhabited by them; and, in consideration of their friendly disposition, and past services to the United States, it is, therefore stipulated, between the United States and the aforesaid Chiefs, that the following reservations shall be surveyed, and marked by the Commissioner, or Commissioners, to be appointed under the 8th article of this; Treaty: For the use of Nea Mathla and his connections, two miles square, embracing the Tuphulga village, on the waters of Rocky Comfort Creek.

For Blunt and Tuski Hajo. a reservation, commencing on the Apalachicola. one mile below Tuski Hajo's improvements, running up said river four miles; thence. west, two miles; thence, southerly to a point two miles due west of the beginning; thence, east, to the beginning point. For Mulatto King and Emathlochee, a reservation. commencing on the Apalachicola. at a point to include Yellow Hair's improvements thence, up said river, for four miles; thence west one mile; thence southerly, to a point one mile west of the beginning; and thence east, to the beginning point. For Econchatimico, a reservation, commencing on the Chatahoochie, one mile below Econchatimico's house; thence, up said river, for four miles; thence, one mile, west; thence, southerly, to a point one mile west of the beginning; thence, east, to the beginning point. The United States promise to guaranty the peaceable possession *of* the said reservations, as defined, to the aforesaid chiefs and their descendants *only, so* long as they shall continue to occupy. improve, or cultivate, the same; but in the event of the abandonment of all, or either of the reservations, by the chief or chiefs, to whom they have been allotted, the reservation, or reservations, so abandoned, shall revert to the United States, as included in the cession made in the first article of this treaty. It is further understood, that the names of the individuals remaining on the reservations aforesaid, shall be furnished, by the chiefs in whose favor the reservations have been made, to the Superintendent or agent of Indian Affairs, in the territory of Florida; and that no other individuals shall be received or permitted to remain within said reservations, without the previous consent *of* the Superintendent or Agent aforesaid; And, as the aforesaid Chiefs are authorized to select the individuals remaining with them, *so* they shall each be separately held responsible *for* the peaceable conduct *of* their towns, or the individuals residing on the reservations allotted them. It is further understood, between the parties, that this agreement is *not* intended to prohibit the voluntary removal, at any future period, of all or either *of* the aforesaid Chiefs. and their connections, *to* the district of country south, allotted to the Florida Indians, by the second article *of* this Treaty, whenever either, or all may think proper to make such an election; the United States reserving the right of ordering, *for* any outrage or misconduct, the aforesaid Chiefs, or either of them, with their connections, within the district *of* country south, aforesaid. It is further stipulated, by the United States, that, of the six thousand dollars, appropriated for implements of husbandry, stock, &c. in the third article of this Treaty, eight hundred dollars shall be distributed, in the same manner, among the aforesaid chiefs and their towns; and it is understood. that, *of* the annual sum *of* five thousand dollars, *to* be distributed by the President *of* the United States, they will receive their proportion. It is further stipulated, that, *of* the *four* thousand five hundred dollars, and two thousand dollars, provided *for* by the 5th article of this Treaty, for the payment for improvements and transportation, five hundred dollars shall be awarded to Neo Mathla, as a compensation *for* the improvements abandoned by him, as well as to meet the expenses he will unavoidably be exposed to, by his own removal, and that *of* his connections.

In testimony whereof, the commissioners, William P. Duval, James Gadsden, and Bernard Segui. and the undersigned chiefs and warriors, have hereunto subscribed their names and affixed their seals. Done at camp, on Moultrie creek, in the territory of Florida, this eighteenth day of September, one thousand eight hundred and twenty-three, and of the independence of the United States the forty-eighth.

> Wm. P. Duval, his x mark, [L. s.]
> James Gadsden, [L. s.]
> Bernard Segui, [L. s.]
> Nea Mlithla, his x mark, [L. s.]
> John Blunt, his x mark, [L. s.]
> Tuski Hajo, his x mark, [L. s.]
> Mulatto King, his x mark, [L. s.]
> Emathlochee, his x mark [L. s.]
> Econchatimico, his x mark [L. s.]

(Author's note: the United States failed to guarantee "peaceable possession" as stipulated in the Treaty of Moultrie Creek.)

Appendix III

Treaty with the Seminole (Treaty of Payne's Landing)

UNITED STATES STATUTES AT LARGE 7: 368-369, 1832.

The Seminole Indians, regarding with just respect, the solicitude manifested by the President of the United States for the improvement of their condition, by recommending a removal to a country more suitable to their habits and wants than the one they at present occupy in the Territory of Florida, are willing that their confidential chiefs, Jumper, Fuch-a-lus-ti-had-jo, Charley Emartla, Coi-had-jo, Holati-Emartla, Ya-hadjo, Sam Jones, accompanied by their agent Major Phagan, and their faithful interpreter Abraham, should be sent at the expense of the United States as early as convenient to examine the country assigned to the Creeks west of the Mississippi river, and should they be satisfied with the character of that country, and of the favorable disposition of the Creeks to reunite with the Seminoles as one people; the articles of the compact and agreement, herein stipulated at Payne's landing on the Ocklewaha river, this ninth day of May, one thousand eight hundred and thirty-two between James Gadsden, for and in behalf of the Government of the United States, and the undersigned chiefs and head-men for and in behalf of the Seminole Indians, shall be binding on the respective parties.

ARTICLE I. The Seminole Indians relinquish to the united States, all claim to the lands they at present occupy in the Territory of Florida, and agree to emigrate to the country assigned to the Creeks, west of the Mississippi river; it being understood that an additional extent of territory, proportioned to their numbers, will be added to the Creek country, and that the Seminoles will be received as a constituent part of the Creek nation, and be re-admitted to all the privileges as members of the same.

ARTICLE II. For and in consideration of the relinquishment of claim in the first article of this agreement, and in full compensation for all the improvements, which may have been made on the lands thereby ceded; the United States stipulate to pay to the Seminole Indians, fifteen thousand, four hundred (15.400) dollars, to be divided among the chiefs and warriors of the several towns, in a ratio proportioned to their population, the respective proportions of each to be paid on their arrival in the country they consent to remove to; it being under stood that their faithful interpreters Abraham and Cudjo shall receive two hundred dollars each of the above sum, in full remuneration for the improvements to be abandoned on the lands now cultivated by them.

ARTICLE III. The United States agree to distribute as they arrive at their new homes in the Creek Territory, west of the Mississippi river, a blanket and a homespun frock, to each of the warriors, women and children of the Seminole tribe of Indians.

ARTICLE IV, The United States agree to extend the annuity for the support of a blacksmith, provided for in the sixth article of the treaty at Camp Moultrie for ten (10) years beyond the period therein stipulated, and in addition to the other annuities secured under that treaty: the United States agree to pay the sum of three thousand (3,000) dollars a year for fifteen (15) years, commencing after the removal of the whole tribe; these sums to be added to the Creek annuities. and the whole amount to be so divided. that the chiefs And warriors of the Seminole Indians may receive their equitable proportion of the same as members of the Creek confederation-

ARTICLE V. The United States will take the cattle belonging to the Seminoles at the valuation of some discreet person to be appointed by the President, and the same shall be paid for in money to the respective owners, after their arrival at their new homes; or other cattle such may be desired will be furnished them, notice being given through their agent of their wishes upon this subject, before their removal, that time may be afforded to supply the demand.

ARTICLE VI. The Seminoles being anxious to be relieved from repeated vexatious demands for slaves and other property, alleged to have been stolen and destroyed by them, so that they may remove un embarrassed to their new homes; the United States stipulate to have the same property investigated, and to liquidate such as may be satisfactorily established, provided the amount does not exceed seven thousand dollars.-

ARTICLE VII. The Seminole Indians will remove within three (3) years after the ratification of this agreement, and the expenses of their removal shall be defrayed by the United States, and such subsistence shall also be furnished them for a term not exceeding twelve (12) months, after their arrival at their new residence; as in the opinion *of* the President, their numbers and circumstances may require, the emigration *to* commence as early as practicable in the year eighteen hundred and thirty-three (1833). and with those Indians at present occupying the Big Swamp, and other parts *of* the country beyond the limits as defined in the second article of the treaty concluded at Camp Moultrie creek, so that the whole *of* that proportion *of* the Seminoles may be removed within the year aforesaid, and the remainder *of* the tribe, in about equal proportions, during the subsequent years *of* eighteen hundred and thirty-four and five, (1834 and 1835.)-

In testimony whereof, the commissioner, James Gadsden, and the undersigned chiefs and head men *of* the Seminole Indians, have hereunto subscribed their names and affixed their seals. Done at camp at Payne's landing, on the Ocklawaha river in the territory of Florida, on this ninth day *of* May, one thousand eight hundred and thirty-two, and of the independence *of* the United States *of* America the fifty-sixth.

James Gadsden, [L. S.]
Holati Ewartla, his x mark, [L. s.]
Jumper, his x mark, [L. S.]
Fuch-ta-Ius-ta-Hadjo, his x mark, [L. s.]
Charley Emartla, his x mark, [L. s.]
Coa Hadjo, his x mark, [L. s.]
Ar-pi-uck-i, or Sam Jones, his x mark, [L. s.]
Ya-ha Hadjo, his x mark, [L. s.]
Mico-Noba, his x mark, [L. s.]

Tokose-Emartla, or Jno. Hicks, his x mark, [L. s.]
Cat-sha-Tusta-nuck-i, his x mark, [L. s.]
Hola-at-a-Mico, his mark, [L. s.]
Hitch-it-i-Mico, his x mark, [L. s.]
E-ne-hah, his x mark, [L. s.]
Ya-ha-emartla Chup-ko, his x mark [L. s.]
Moke-his-she-lar-ni, his x mark, [L. s.]

Witnesses:

Douglas Vass, Secretary to Commissioner,
John Phagan, Agent,
Stephen Richards,
Abraham, Interpreter, his x mark

Cudjo, Interpreter, his x mark
Erastus Rogers,
Interpreter B. Joscan

(Author's note: Known colloquially as "The Treaty of Payne's Landing" it was ineffective, as the United States failed in its promise of "peaceable possession," and it was broken when the Indians were forcibly removed to Indian Territory west of the Mississippi a decade later regardless of their "satisfaction with the character of that country." The Treaties of Moultrie Creek and Payne's Landing are the only ones ever signed by members of Florida's Seminole and Miccosukee Tribes.)

Appendix IV

Metropolitan Miami–Dade County Historic Preservation Ordinance

SEC. 16A-2. DECLARATION OF LEGISLATIVE INTENT.

It is hereby declared as a matter of public policy that the protection, enhancement and perpetuation of properties of historical, cultural, archeological, paleontological, aesthetic and architectural merit *are* in the interests of the health, prosperity and welfare of the people of Miami–Dade County. Therefore, this chapter is intended to:

(1) Effect and accomplish the protection, enhancement and perpetuation of buildings, structures, improvements, landscape features, paleontological and archeological resources of sites and districts which represent distinctive elements of the County's cultural, social, economic, political, scientific, religious, prehistoric and architectural history;

(2) Safeguard the County's historical, cultural, archeological, paleontological and architectural heritage, as embodied and reflected in such individual sites, districts and archeological zones;

(3) Foster civic pride in the accomplishments of the past;

(4) Protect and enhance the County's attraction to visitors and the support and stimulus to the economy thereby provided; and

(5) Promote the use of individual sites and districts for the education, pleasure and welfare of the people of Miami–Dade County.

(Ord. No. 81-13, § 2, 2-17-81; Ord. No. 03-38, § 1, 3-11-03)

Sec. 16A-3. Scope of regulations.

(1) This chapter is intended to and shall govern incorporated and unincorporated Miami–Dade County.

(2) The regulatory jurisdiction of the Miami–Dade County Historic Preservation Board pursuant to this Chapter shall extend to:
- (a) all property located in the unincorporated *areas* of Miami–Dade County;
- (b) all property located in incorporated *areas* of Miami–Dade County except where the municipality has enacted its own historic preservation ordinance in accordance with section 16A-3.1;
- (c) archeology and paleontology zones and sites in the incorporated and unincorporated *areas* of Miami–Dade County except where the municipality has enacted its own historic preservation ordinance in accordance with section ,16A-3.1 and the municipality, within 365 days of the effective date of this ordinance, enacts an ordinance that (1) expressly retains jurisdiction over archeology and paleontology zones and sites, (2) adopts regulations as least as protective of archeology and paleontology zones and sites as those in this Chapter, and (3) commits the municipality to retain sufficient archeological personnel *or* consultants to enforce such regulations; and
- (d) the enforcement of the minimum standards established by this Chapter as set forth in this Chapter.

(3) Nothing contained herein shall be deemed to supersede *or* conflict with applicable building and zoning codes. Provisions contained herein shall be cumulative and read in conjunction with other provisions of the Miami–Dade County Code.

(Ord. No. 81-13, § 3, 2-17-81; Ord. No. 82-99, § 1,10-19-82; Ord. No. 03-38, § 2,3-11-0

Sec. 16A-4. Definitions.

(1) *Archeological or paleontological zone:* An area designated by this chapter which is likely to yield information on the paleontology, history and prehistory of Miami–Dade County based on prehistoric settlement patterns in Miami–Dade County as determined by the results of the Miami–Dade County historic survey. These zones will tend to conform to natural physiographic features which were the focal points for prehistoric and historic activities and paleontology.

(2) *Certificate of appropriateness:* A certificate issued by the Board permitting certain alterations or improvements to a designated individual site or property in a designated district.

    (a) *Regular certificate of appropriateness:* A regular certificate of appropriateness shall be issued by the staff of the Preservation Board, based on the guidelines for preservation approved by the Board.

    (b) *Special certificate of appropriateness.* For all applications for a special certificate of appropriateness involving the demolition, removal, reconstruction or new construction at an individual site or in a district, a special certificate of appropriateness is required that is issued directly by the Board.

(3) *Certificate to dig:* A certificate that gives the Board's permission for certain digging projects that may involve the discovery of as yet unknown or known archeological or paleontological sites in an archeological or paleontological zone. This certificate is issued by staff of the Board based on the guidelines for preservation approved by the Board.

(4) *Certificate of recognition:* A certificate issued by the Board recognizing properties designated pursuant to this chapter.

(5) *Demolition:* The complete constructive removal of a building on any site.

(6) *Districts:* A collection of archeological or paleontological sites, buildings, structures, landscape features or other improvements that are concentrated in the same area and have been designated as a district pursuant to this chapter.

(7) *Exterior:* All outside surfaces of a building or structure.

(8) *Guidelines for preservation:* Criteria established by the Preservation Board to be used by staff in determining the validity of applications for a regular certificate of appropriateness and any certificate to dig and to establish a set of guidelines for the preservation of buildings in south Florida.

(9) *Historic Preservation Board:* A board of citizens created by this chapter as described in Sections 16A-5 through 16A-9.

(10) *Historic survey:* A comprehensive survey compiled by the Historic Preservation Division of the Miami–Dade County Office of Community and Economic Development involving the identification, research and documentation of buildings, sites and structures of any historical, cultural, archeological, paleontological or architectural importance in Miami–Dade County, Florida.

(11) *Individual site:* An archeological site, a paleontological site, building, structure, place or other improvement that has been designated

as an individual site pursuant to this chapter. Under the provisions of this chapter interior spaces may be regulated only where a building or structure is a designated individual site and where its interiors are specifically designated.

(12) *National Register* of *Historic Places:* A federal listing maintained by the U.S. Department of the Interior of buildings, sites, structures and districts that have attained a quality of significance as determined by the Historic Preservation Act of 1966 as amended.

(13) *Ordinary repairs or maintenance:* Work done to prevent deterioration of a building or structure or decay of or damage to a building or structure or any part thereof by restoring the building or structure as nearly as practicable to its condition prior to such deterioration, decay or damage..

(14) *Owner* of a *designated property:* As reflected on the current Metropolitan Miami–Dade County tax rolls or current title holder.

(15) *Undue economic hardship:* Failure to issue a certificate would place an onerous and excessive financial burden upon the owner that would amount to the taking of the owner's property without just compensation.

(16) *Landscape feature:* Any improvement or vegetation including, but not limited to outbuildings, walls, courtyards, fences, shrubbery, trees, sidewalks, planters, plantings, gates, street furniture and exterior lighting.

(Ord. No. 81-13, § 4,2-17-81; Ord. No. 82-99, § 1, 10-19-82; Ord. No. 03-38, § 5, 3-11-03)

Sec. 16A-14. Certificates to dig.

(I) Within an archeological or paleontological zone, new construction, filling, digging, the removal of trees or any other activity that may alter or reveal an interred archeological or paleontological site shall be prohibited without a certificate All applications to all appropriate municipal or County agencies involving new construction, large-scale digging, the removal of trees or any other activity that may reveal or disturb an interred archeological or paleontological site, in an archaeological or paleontological zone shall require a certificate to dig before approval. Based on the designation report for the archeological or paleontological zone, a complete application for a certificate to dig and any additional guidelines the Board may deem necessary, the staff of the Board shall, within ten (10) days from the date the completed application has been filed, approve the application for a certificate to dig by the owners of a property in a designated archeological or paleontological zone. The certificate to dig may be made subject to specified conditions, including but not limited to conditions regarding site excavation. In order to comply with the site excavation requirements of the certificate to dig, the applicant may agree to permit the County Archeologist to conduct excavation from the time of the approval of the certificate to dig until the effective date thereof. The findings of the staff shall be mailed to the applicant by registered mail promptly. The applicant shall have the opportunity to challenge the staff decision or any conditions attached to the certificate to dig by requesting a meeting of the Board. The Board shall convene within thirty-five (35) days after such a request and shall make every effort to review and reconsider the original staff decision to arrive at an equitable decision. The decision of the Board shall be reduced to writing within seven (7) days from the date of the meeting.

(II) *Approved certificates to dig.* Approved certificates to dig shall contain an effective date not to exceed sixty (60) days at .which time the proposed activity may begin, unless the Board decides to designate the site in question as an individual site or district pursuant to Section 16A-10 in which all the rules and regulations pertaining to the designation process shall apply from the date the designation report has been filed.

(III) *[Work to conform to certificate; stop walk order.]* All work performed pursuant to the issuance of a certificate to dig shall conform to the requirements of such certificate. It shall be the duty of the appropriate government agencies and the staff of the .Board to inspect from time to time any work pursuant to such certificate to assure compliance. In the event work is performed not in accordance with such certificate, the official designated by the County Manager pursuant to Section 16A-11 (IX)[VIII] shall be empowered to issue a stop work order and all work shall cease. No person, firm or corporation shall undertake any work on such projects as long as such stop work order shall continue in effect.

[Complete Miami–Dade Ordinances are available on line at http://livepublish.municode.com/4/lpext.dill?f=templates&fn=main-j.htm&vid=10620]

Appendix V
Federal and State of Florida Statutes Relating to Archaeological Investigations

Organizations, firms, and/or individuals planning to conduct archaeological investigations must be familiar with key statutes and rules that are likely to affect their activities. It is incumbent upon them to obtain appropriate sets of the regulations from the Division of Historical Resources (DHR) in Tallahassee.

FEDERAL ACTS

1906 — Antiquities Act — requires permits for excavation and imposes penalties for damaging historic or prehistoric ruins on federally owned land, as amended.

1935 — Historic Sites Act — launched the National Historic Landmarks program, as amended.

1960 — Reservoir Salvage Act — established procedures for conducting archaeological investigations and salvage before construction of federal dams and reservoirs.

1966 — National Historic Preservation Act — directs the Secretary of the Interior, through the National Park Service, to maintain and expand the National Register of Historic Places, including buildings, sites, districts, structures, and objects significant in American ... archaeology and culture, as amended.

1974 — Reservoir Salvage Act — broadened to apply to all federal projects, as amended.

1979 — The Archaeological Resources Protection Act — regulates removal of archaeological resources from federal land, as amended.

1983 — Convention on Cultural Property Implementation Act — prohibits entry of looted antiquities into the United States if requested by another country.

1987 — The Abandoned Shipwreck Act — removes abandoned historic shipwrecks in U.S. waters from the jurisdiction of admiralty courts and transfers title to the states.

1990* — Native American Graves Protection and Repatriation Act — requires federal agencies and museums to consult with and possibly repatriate to Indian tribes Native American human remains and sacred objects, as amended.

A list of Laws, Regs, & Standards is available on the National Park Service web site at http://www.cr.nps.gov/linklaws.htm..

The Executive Order regarding Indian Sacred Sites is at http://www.cr.nps.gov/local-law/eol3007.htm.

Executive Order No. 11593, Protection And Enhancement Of The Cultural Environment — http://archnet.asu.edu/archnet/tropical/crm/usdocs/execord.htm.

Archaeology And Historic Preservation: Secretary of the Interior's Standards for Archaeological Documentation —

State of Florida

1993* — Chapter 267 — Historical Resources Act — State policy relative to historic properties:

1. provide leadership in the preservation of the state's historic resources;

2. administer state-owned or state-controlled historic resources in a spirit of stewardship and trusteeship;

3. contribute to the preservation of non-state-owned historic resources and to give encouragement to organizations and individuals undertaking preservation by private means;

4. foster conditions, using measures that include financial and technical assistance, for the harmonious coexistence of society and state historic resources;

5. encourage the public and private preservation and utilization of elements of the state's historically built environment; and

6. assist local governments to expand their historic preservation programs and activities.

It is further declared to be the public policy of the state that all treasure trove, artifacts, and such objects having intrinsic or historical and archaeological value which have been abandoned on state-owned lands or state-owned sovereignty submerged lands shall belong to the state with the title thereto vested in the Division of Historical Resources of the Department of State for purposes of administration and protection.

Rule 1A-32*—Florida Administrative Code—The Archaeological Research Permit Rule—a permit from DHR is required for investigations proposed for state-owned or controlled lands, including sovereignty submerged lands. Additional permits may be required from other agencies, such as dredge and fill for underwater excavation.

Rule 1A-46*—Florida Administrative Code—pertains to the manner in which DHR staff determines the completeness and sufficiency of reports prepared to satisfy environmental review, archaeological research, and grant project requirements.

Chapter 872*—Florida Statutes—Offenses Concerning Dead Bodies and Graves—regards dealing with dead bodies, injuring or removing tombs or monuments and disturbing contents of graves or tombs, cremating human bodies, autopsies, and procedures to follow for discovery of unmarked human burials.

Rule 1A-44*—Florida Administrative Code—applies to discoveries of unmarked human remains when they come under the jurisdiction of the State Archaeologist.

For definitive information, especially regarding those regulations marked with an asterisk, contact:

James J. Miller, State Archaeologist & Chief
Bureau of Archaeological Research
Florida Department of State, Division of Historical Resources,
R.A. Gray Building, 500 South Bornough, Tallahassee, Florida 32399-0250

Note: additional County and local regulations may apply.

[The Florida Division of Historic Resources is found at http://flheritage.com

[Cultural Resource Protection Laws and Regulations at http://dhr.dos.state.fl.us/statutes/index.cfm

[Cultural Resource Protection For Private Landowners—http://www.flheritage.com/culturalmgmt/]

Bibliography

American State Papers [ASP]. *Documents, Legislative and Executive, of the Congress of the United States.* Vol. 1 (March 3, 1789 to March 3, 1819). Washington, D.C.: Gates and Seaton, 1789–1819.
Abbott, R. Tucker. *American Seashells.* 2nd ed. New York: Van Nostrand Reinhold, 1974.
Baier, Elizabeth. "Land May Emerge from Limbo." *Miami Herald,* October 5, 2003, Northwest Section, 3.
Barbour, Thomas. *That Vanishing Eden.* Boston: Little, Brown, 1944.
Bartram, William. *Travels through North and South Carolina, Georgia, East and West Florida, the Cherokee Country, the Extensive Territories of the Muscogulges or Creek Confederacy, and the Country of the Choctaws.* Philadelphia: 1791; London: 1792.
Beard, Daniel. (1938). *Wildlife Reconnaissance.* Everglades National Park Project, U. S. Dept. of the Interior, National Park Service. Washington, D.C.. GPO, 1976.
Beriault, John G., Robert S. Carr, Jerry J. Stipp, Richard Johnson and Jack Meeder. "The Archaeological Salvage of the Bay West Site, Collier County, Florida." *The Florida Anthropologist* 34:2 (1981), 39–58.
Blake, Nelson M. *Land into Water—Water into Land.* Gainesville: University Press of Florida, 1980.
Blanchard, Charles E. *New Worlds, Old Songs.* Gainesville: Institute of Archaeology and Paleoenvironmental Studies, University of Florida, 1995.
Bower, B. "Miami Ice Age Site Yields Rich Haul." *Science News* (Jan. 25, 1986), 129:52.
_____. "Amazon Cave Yields Ancient Culture." *Science News* (Apr. 20, 1996), 149:16, 244.
Brooks, H. Kelly. "Lake Okeechobee." *Environments in South Florida: Present and Past,* ed. P. J. Gleason. Miami Geological Society Memoir 2, rev. ed., Coral Gables, FL, (1984), 38–68.
Brown, Robin C. *Florida's Fossils.* Sarasota, FL: Pineapple Press, 1988.
_____. *Florida's First People: 12,000 Years of Human History.* Sarasota, FL: Pineapple Press, 1994.
Bullen, Adelaid K. *Florida Indians of Past and Present.* Gainesville: University Press of Florida, 1965.
Bullen, Ripley P. *A Guide to the Identification of Florida Projectile Points.* Gainesville: Kendall Books, 1975.
_____, and H. K. Brooks. "Two Ancient Florida Dugout Canoes" *Quarterly Journal of the Florida Academy of Science,* 30:2 (1967), 97–107.
Burgess, Robert F. *The Cave Divers.* New York: Dodd, Mead, 1976.

_____. *Man: 12,000 Years Under the Sea.* New York: Dodd, Mead, 1980.
Canby, Thomas Y. "The Search for the First Americans." *National Geographic,* 156:3 (1979), 330–363.
Carr, R. S., and J. G. Beriault. "Prehistoric Man in Southern Florida." *Environments of South. Florida, Present and Past II,* Ed. P. J. Gleason. Coral Gables, FL: Miami Geological Society, Coral Gables, FL, 1984, 1–14.
_____ and _____. "Prehistoric Man in Southern Florida." Community and Economic Development, Metropolitan Dade County, Historic Preservation Division, 1987, 12.
Carr, Robert S. "Prehistoric Circular Earthworks in South Florida." *The Florida Anthropologist,* 38:4 (1985), 288–301.
_____. "Preliminary Report of Archaeological Excavations at the Cutler Fossil Site in South Florida." 51st Annual Meeting of the Society for American Archaeology. New Orleans, La., April 27, 1986.
_____. "Early Man in South Florida." *Archaeology* (Nov./Dec. 1987), 62–63.
Carr, Robert S., David Dickel and Marilyn Mason. "Archaeological Investigations at the Ortona Earthworks and Mounds." *The Florida Anthropologist* (Dec. 1995), 48:4.
Carr, Robert S., M. Yasar Iscam and Richard Johnson. "A Late Archaic Cemetery in South Florida." *The Florida Anthropologist* 37:4 (1984), 172–188.
Carr, Robert S., and John Ricisak. "The Miami Circle: Beneath the Modern City." *St. Thomas Law Review* 13:1 (2000), 225–232.
Clark, James A., and Craig S. Lingle. "Predicted Relative Sea Level Changes (18,000 years B.P. to Present) Caused by Late Glacial Retreat of the Antarctic Ice Sheet." *Quaternary Research* 11 (1979), 279–298.
Clausen, Carl, A. D. Cohen, Cesare Emiliani, J. A. Holman and J. J. Stipp. "Little Salt Spring, Florida a Unique Underwater Site." *Science* 203 (1979), 609–614.
Clausen, Carl J., H. K. Brooks and A. B. Wesolowsky. "The Early Man Site at Warm Mineral Springs, Florida." *Journal of Field Archaeology* 2 (1975a), 191–213.
_____. "Florida Spring Confirmed as 10,000 Year Old Early Man Site." *The Florida Anthropologist* 38 (1975b).
Cobbs, John L., Flowers, C., Gardner, J. L., Loomis, H. B., eds. *Through Indian Eyes.* The Reader's Digest Association, (1995).
Cocking, Susan. "Underwater time capsule." *The Miami Herald,* June 29, 2003, 17C.
Colburn, David ed. *The African American Heritage of Florida.* Gainesville: University Press of Florida, 1995.
Cook, Michael. *A Brief History of the Human Race.* New York: W.W. Norton, 2003.
Covington, James W. "Migration of the Seminoles into Florida, 1700–1820." *Florida Historical Quarterly* 46 (1968), 340–344.
_____. *The Seminoles of Florida.* Gainesville: University Press of Florida, 1993.
Crankshaw, Joe. "Early State History Offers Bloody Lesson in Intolerance." *The Miami Herald,* May 29, 1995.
Cushing, Frank Hamilton. "Exploration of Ancient Key Dwellers' Remains on the Gulf Coast of Florida." *Proceedings of the American Philosophical Society* 35:153 (1897), 329–448.
Davis, Frederick T. "History of Juan Ponce de Leon's Voyage to Florida." *Florida Historical Quarterly* (Florida Historical Society) XIV (July 1935), 1, 16–23.
DePrather, Chester B. *Late Prehistoric and Early Historic Chiefdoms in the Southeastern United States.* New York: Garland Press, 1991.
Derr, Mark. *Some Kind of Paradise.* New York: William Morrow, 1989.
Dewar, Elaine. *Bones: Discovering The First Americans.* New York: Carroll & Graf, 2001.
Dickinson, Jonathan. *Jonathan Dickinson's Journal or God's Protecting Providence,* ed.

Evangeline Walker Andrews and Charles McLean Andrews. Stuart, FL: Valentine Books, 1975.
Dolan, Edward M. "Florida's Relationship to the Antilles and Mesoamerica: A Synthesis." *Southern Indian Studies* XI (Oct. 1959), 3–15.
Doran, G. H., D. N. Dickel, and L. A. Newsom. "A 7,290-Year-Old Bottle Gourd from the Windover Site, Florida." *American Antiquity* 55 (1990), 354–360.
Dormer, Elinore M. *The Shell Islands*. New York Vantage Press, 1975.
Douglas, Marjorie Stoneman. *Voice of the River*. Sarasota, FL: Pineapple Press, 1990.
Dunbar, James S., and Ben I. Waller. "A Distribution Analysis of the Clovis/Suwannee Paleo-Indian Sites of Florida—A Stratigraphic Approach." *The Florida Anthropologist* 36:1–2 (1983), 18–30.
Eaton, S. Boyd, Marjorie Shostak and Melvin Konner. *The Paleolithic Prescription*. New York: Harper & Row, 1988.
Ennemoser, Rusty Sevigny. "Florida's Prehistoric Monuments." *Florida Heritage*. Division of Historical Resources, Florida Department of State, 1993, 12–13.
Eyster, Irving R. "Excavations of the Arch Creek Mill Site." Report on file, Dade County Historic Preservation Division, 1981, 8.
_____. "Field Notes, 8 MO 2, Stock Island, Key West, FL."
_____. *Handbook of South Florida Archaeology*. Paper.
Fagan, Brian M. *The Great Journey*. New York: Thames and Hudson, 1987.
_____. *Ancient North America*. New York: Thames & Hudson, 1991.
Fairbanks, Charles. "Creek and Pre-Creek." *Archaeology of the Eastern United States*, ed. J. B. Griffin. Chicago: University of Chicago Press, 1952, 285–300.
_____. "The Function of Black Drink Among the Creeks." in *Black Drink: A Native American Tea*, ed. Charles M. Hudson. Athens: University of Georgia Press, 1979, 120–149.
Felmley, Amy. "Osteological Analysis of the Pine Island Site Human Remains." *The Florida Anthropologist* 43:4 (1990), 262–274.
Fontaneda, Do. D'Escalante. *Memoir of Do. D'Escalante Respecting Florida Translated by Buckingham Smith, Washington, 1854*. Reprinted with revisions, University of Miami Press and the Historical Association of Southern Florida, Misc. Publ. No.1, 1944.
"Forest buried by Time and Sea." *The Island Packet* Hilton Head, S.C., October 28, 1994.
Fritchie, John. *Everglades Journal*. Miami: Florida Heritage Press, 1992.
Gaby, Donald C. *The Miami River and Its Tributaries*. The Historical Association of Southern Florida, 1994.
George, Paul, ed. *A Guide to the History of Florida*. Westport, CT: Greenwood Press, 1989.
Gidwitz, Tom. "Telling Time." *Archaeology* 54:2 (Mar./Apr. 2001), 36–41.
Gilliland, Marion S. *Key Marco's Buried Treasure*. Gainesville: University of Florida Press, 1975.
_____. *The Calusa Indians of Florida*. Gainesville: University of Florida Press, 1996.
Gleason, P. J., ed. *Environments of South Florida: Past and Present*. Miami Geological Society Memoir 2, revised edition, 1984.
Goggin, John M. "Archaeological Investigations on the Upper Florida Keys." *Tequesta*. 1:4 (1944), 13–35.
_____. *Indian and Spanish Selected Writings*. Coral Gables, FL: University of Miami Press, 1964.
_____. "The Indians and History of the Matecumbe Region." *Tequesta* 10 (1950), 13–24.
_____. "The Snapper Creek Site." *The Florida Anthropologist* 3:3–4 (1950), 50–64.
Gore, Rick. "The First Europeans." *National Geographic* 192:1 (1997), 96–113.

Bibliography

———. "The Most Ancient Americans." *National Geographic* 192:4 (Oct. 1997), 92–99.

Green, Rayna, and Melanie Fernandez. *The British Museum Encyclopedia of Native North America*. Bloomington: Indiana University Press, 1999.

Griffin, John W. *The Archeology of Everglades National Park: A Synthesis*. National Park Service Southeast Archeological Center, Tallahassee, FL: 1988.

———. *Archaeology of the Everglades*. Gainesville: University Press of Florida, 2003.

———. *Fifty Years of Southeastern Archaeology*, ed. Patricia Griffin. Gainesville: University Press of Florida, 1996.

Griffin, John W., Mildred L. Fryman and James J. Miller. *Cultural Resource Reconnaissance of the National Key Deer Wildlife Refuge*. Tallahassee, FL: Interagency Archaeological Services—Atlanta for U.S. Fish And Wildlife Service by Cultural Resource Management, Inc., 1979.

Griffin, et al. "Excavations at the Granada Site." *Archaeology and History of the Granada Site*, Vol. I, Div. Of Archives, History and Records Management, Florida Dept. of State, 1984.

Griffin, John W., Sue B. Richardson, Mary Pohl, Carl D. McMurray, C. Margaret Scarry, Suzanne K. Fish, Elizabeth S. Wing, L. Jill Grimm, E. C., G. L. Jacobson, Jr., W. A. Watts, B. C. S. Hansen, and K. A. Maasch. "A 50,000-Year Record of Climate Oscillations from Florida and Its Temporal Correlation with the Heinrich Events." *Science* 261 (1993), 198–200.

Gorman, John. "The Secrets of the Spring." *Aloft* 11:8 (1979), 6–13.

Grun, Bernard. *The Timetables of History*. New York: Simon and Schuster, 1987.

Haeussler, A.M., and D.H. Morris. "Human Mandible from Warm Mineral Spring." *American Journal of Physical Anthropology* 81:2 (1990), 233–234.

Haiduven, Richard. "Miami Circle Fact Sheet." *The Archaeological Society of Southern Florida* 25:5 (May 1999), 2.

Hammond, E.A. "Dr. Strobel Reports on Southeast Florida, 1836." *Tequesta* (Journal of the Historical Association of Southern Florida) XXI (1961), 69.

Hann, John M. *Missions to the Calusa*. Gainesville: University of Florida Press, 1991.

Hartman, Todd. "Housing May Circle an Indian Burial Site." *The Miami Herald*, Jan. 26, 1995, 1B.

Haynes, Gary. *The Early Settlement of North America*. Port Chester, NY: University of Cambridge Press, 2002.

Hoffmeister, John Edward. *Land from the Sea*. Coral Gables, FL: University of Miami Press, 1974.

Hothem, Lar. *Arrowheads and Projectile Points*. Paducah, KY: Collector Books, 1991.

Hrdlicka, Ales. *The Anthropology of Florida*. Florida State Historical Society, 1922.

Hudson, Charles. *The Southeastern Indians*. Knoxville: University of Tennessee Press, 1976.

———, and Carmen C. Tesser. *The Forgotten Centuries: Indians and Europeans in the American South, 1521–1704*. Athens: University of Georgia Press, 1994.

Iscan, M. Yasar. "Skeletal Biology of the Margate-Blount Population." *The Florida Anthropologist* 36:3-4 (1983), 154–166.

Iscan, Mehmet Yasar, Morton H. Kessel and Robert S. Carr. "Human Remains from the Archaic Brickell Bluff Site." *The Florida Anthropologist* 46:4 (1993), 277–281.

Ingold, Tim, Ed. *Companion Encyclopedia of Anthropology*. New York: Routledge, 1994, xiii-xiv.

Job, Herbert K. "In the Cape Sable Wilderness," *Tales of Old Florida*, ed. Frank Oppel and Tony Meisel. Secaucus, NJ: Castle, 1903, 139.

Kenworthy, Charles J. "Ancient Canals in Florida." *Smithsonian Institution Annual Report*. Washington, D.C.: GPO, 1881, 631–635.

Klingener, Nancy. "Rise in Ocean Levels Poses Real Threat, Keys Group Told." *The Miami Herald*, May 28, 1999, 5B.
Kozuch, Laura. *Sharks and Shark Products in Prehistoric Florida*. Institute of Archaeology and Paleoenvironmental Studies, Gainesville: University of Florida, 1993.
Lamme, Vernon. "The Arch Creek Site, Dade County." *Florida Anthropologist* 28:1 (1975), 1-13.
____. *Florida Lore Not Found in History Books*. Boynton Beach, FL: Star Publishing, 1973, 76-85, 115-116.
Land, Douglas. "Highlights and Observations, Clovis and Beyond Conference." *Indian Artifact Magazine* 19:1 (2000), 52-53, 65.
Larson, Lewis H. "Aboriginal Subsistence Technology on the Southeastern Coastal Plain During the Late Prehistoric Period." *Ripley P. Bullen Monographs in Anthropology and History, No. 2*, the Florida State Museum. Gainesville: University Press of Florida, 1980.
Laxson, Dan. "The Arch Creek Site." *The Florida Anthropologist* 10:3-4 (1957), 10.
____. "The DuPont Plaza Site." *The Florida Anthropologist* 21:2-3 (1968), 55-60.
Lemonick, Michael D., and Andrea Dorfman. "Up from the Apes." *Time* 154:8 (Aug. 23, 1999), 51-58.
Lewis, Clifford M. "The Calusa." *Tacachale*, ed. Jerald T. Milanich and Samuel Proctor. Gainesville: University of Florida Press, 1978, 19-58.
Loucks, L., and Marcia K. Welch. "Excavations at the Granada Site." *Archaeology and History of the Granada Site*, Vol. I, Florida Division of Archives, History and Records Management, 1982.
Luer, George. "Calusa Canals in Southwest Florida." *The Florida Anthropologist* 42:2 1989.
____. "Further Research on the Pine Island Canal and Associated Sites, Lee County, Florida." *The Florida Anthropologist* 42:3 (1989), 241-247.
____. "Three Aboriginal Shell Middens On Longboat Key, Florida: Manasota Period Sites of Barrier Island Exploitation." *Florida Anthropologist*, 32 (1979), 34-45.
MacCauley, Clay. *The Seminole Indians of Florida*. Gainesville: University Press of Florida, 2000.
MacNeish, Richard S., Paul C. Mangeldorf and Walton C. Galinat. "Domestication of Corn." *Science* 143.3606 (1964), 538-545.
Marquardt, William. *Culture and Environment in the Domain of the Calusa*. Institute of Archaeological and Paleoenvironmental Studies, Gainesville: University of Florida, 1992.
____. "The Development of Cultural Complexity in Southwest Florida: Elements of a Critique." *Southeastern Archaeology* 5:1 (1986), 63-70.
____. "October, 1697: A Clash of Ideas." *Calusa News* 2:4 (March 1998).
____. "Pineland." *Calusa News* 5:6 (Oct. 1990), 12.
____. "Return to Battey's Landing," *Calusa News* 3:1-3 (May 1989).
____. "A Shell Tool Workshop on Useppa Island." *Calusa News*. Florida Museum of Natural History, Gainesville: University of Florida, 1992, 6:5.
____. "Useppa Island." *Calusa News*. Florida Museum of Natural History. Gainesville: University of Florida, 1990, 5:4-5.
"Mastoson's Gourd Meals Dash Dispersal Theory." *National Geographic* 184:4 (Oct. 1993), 25.
Matthews, Samuel W. "This Changing Earth." *National Geographic* 143:1 (1973), 1-37.
Maxwell, James A., ed. *America's Fascinating Indian Heritage*. The Reader's Digest Association, 1995.
McCarthy, Kevin M. *Native Americans in Florida*. Sarasota, FL: Pineapple Press, 1999.

McDonald, Jerry N., and Susan L. Woodward. *Indian Mounds of the Atlantic Coast: A Guide to Sites from Maine to Florida*. Newark, OH: McDonald and Woodward Publishing Co., 1987.

McEwan, Bonnie G., ed. *Indians of the Greater Southeast*. Gainesville: University Press of Florida, 2000.

McGoun, William E. *Prehistoric Peoples of South Florida*. Tuscaloosa: University of Alabama Press, 1993.

———. *Ancient Miamians, the Tequesta of South Florida*. Gainesville: University Press of Florida, 2002.

McNicoll, Robert E. "The Caloosa Village Tequesta: A Miami of the Sixteenth Century." *Tequesta* (Bulletin of the University of Miami) (1941), 11–20.

Mehringer, Peter J., and Warren Morgan. "Clovis Cache Found: Weapons of Ancient Americans." *National Geographic* (1988), 500–503.

Merzer, Martin. "Secrets of Past Centuries Emerge from Under Miami Parking-lot Site," *The Miami Herald*, Sept. 3, 2003, 1B–4B.

Milanich, J. T., and Susan Milbrath, ed. *First Encounters: Spanish Explorers in the Caribbean and the United States, 1492–1570*. Gainesville: University of Press of Florida, 1989.

Milanich, Jerald T. *Archaeology of Precolumbian Florida*. Gainesville: University Press of Florida, 1994.

———. *Florida Indians and the Invasion from Europe*. Gainesville: University Press of Florida, 1998.

———. *Florida's Indians from Ancient Times to the Present*. Gainesville: University Press of Florida, 1998.

———. "A Host of Would-be Conquerors." *Archaeology* 49:1 (1996), 68–69.

———. "Laboring in the Fields of the Lord." *Archaeology* 49:1 (1996), 60–67.

———. "Osceola's Head." *Archaeology* 57 (2004), 48–53.

Milanich, Jerald T., and Charles H. Fairbanks. *Florida Archaeology*. New York: Academic Press, 1980.

Milanich, Jerald T., and James J. Miller. *Archaeology of the Everglades*. Gainesville: University Press of Florida, 2002.

Mowers, Bert, and Wilma B. Williams. "Cagles Hammock, Coral Springs Site No.5." *The Florida Anthropologist* 27:4 (1974), 171–179.

———. "The Peace Camp Site, Broward County, Florida." *The Florida Anthropologist* 42 (1972), 1–20.

Murray, Marian. *Florida Fossils*. Tampa, FL: Trend House, 1975, 29.

"A New Theory of Prehistoric Migration." *The Miami Herald*, Nov. 1, 1999, 3A.

Newman, Christine L. "The Cheetum Site: An Archaic Burial Site in Dade County Florida." *The Florida Anthropologist* 46:1 (1993), 37–42.

Newman, T. Stell. "Preliminary Historical Studies Plan, Biscayne National Monument, Homestead, FL." Denver Service Center Historic Preservation Team, N.P.S., U.S.D.I., 1975.

Newsom, Lee S. David Webb and James S. Dunbar. "History and Geographic Distribution of *Curcurbita pepo* Gourds in Florida." *Journal of Ethnobiology* 13:1 (1993), 75–97.

Newsom, Lee Ann, Ph.D., Curator. Center for Archaeological Investigations, Southern Illinois University at Carbondale, Personal communications, 1994–1995, 2003.

Newsom, Lee Ann, and Barbara Purdy. "Florida Canoes: A Maritime Heritage from the Past." *The Florida Anthropologist* 43:3 (1990), 164–180.

Newsom, Lee A., and C. Margaret Scary. *Homegardens and Mangrove Swamps: Pineland Archaeobotanical Research*. Unpublished manuscript, 2003.

Olsen, Fred. *On the Trail of the Arawaks*. 1st ed. Norman: University of Oklahoma Press, 1974.
Overstreet, Robert M. *Official Overstreet Identification and Price Guide to Indian Arrowheads*, 3rd. ed., NY: House of Collectibles, 1993.
Paige, John C., and Lawrence F. Van Horn. *An Ethnohistory of Big Cypress National Preserve, Florida*, N.P.S., U.S.D.I, 1982.
____ and ____. *Historic Resource Study for Everglades National Park*. N.P.S., U.S.D.I., 1986.
Parfit, Michael. "Hunt for the First Americans." *National Geographic* 198:6 (Dec. 2000), 41–067.
Parks, Arva Moore. "Where the River Found the Bay." *Archaeology and History of the Granada Site*, Vol. II, Florida Division of Archives, History and Records Management, 1982.
Payne, Claudine. "Horr's Island Yields a New View of the Florida Archaic." *Calusa News* (Florida Museum of Natural History) 6 (1992), 1–2.
Perry, I. Mac. *Indian Mounds You Can Visit*. St. Petersburg, FL: Great Outdoors Publishing Co., 1993.
Pollack, Phillip M. *300' X 35 miles, Corridor to the Past*. A Joint Project of the Florida Dept. of Transportation and Div. of Historical Resources, Florida Department of State, 1986.
Purdy, Barbara A. *The Art and Archaeology of Florida's Wetlands*. Boca Raton, FL: CRC Press, 1991.
____. *Florida's Prehistoric Stone Technology*. Gainesville: University Press of Florida, 1981.
____. *How to Do Archaeology the Right Way*. Gainesville: University Press of Florida, 1996.
____. "The Temporal and Spatial Distribution of Bone Points in the State of Florida." *The Florida Anthropologist* 26 (1973), 143–152.
Renfrew, Colin, and Paul Bahn. *Archaeology: Theories, Methods and Practice*. New York: Thames and Hudson, 1991.
Renz, Mark. *Fossiling in Florida*. Gainesville: University Press of Florida, 1999.
Roberts, William. *An Account of First Discovery and Natural History of Florida, a Facsimile Reproduction of the 1763 Edition*. Gainesville: University Press of Florida, 1976.
Robertson, William D. D. *The History of America*. 1st volume, Book II. Dublin: University of Edinburgh, 1777.
Romans, Bernard. *A Concise Natural History of East and West Florida, a Facsimile Reproduction of the 1775 Edition*, with Introduction by Rembert W. Patrick. Gainesville: University of Florida Press, 1962.
Rouse, Irving. "Vero and Melbourne Man: A Cultural and Chronological Interpretation." *Transactions of the New York Academy of Science* Ser.2, 12:7 (1950), 220–224.
Royal, William, and Eugenie Clark. "Natural Preservation of Human Brain, Warm Mineral Springs, Florida." *American Antiquity* 26, 2, (Oct. 1960), 285–287.
Russo, Michael, Ann Cordell, Lee Newsom and Sylvia Scudder. "Glades Ceramics, Archaeobotanical and Soils Analyses, plus Appendices and References from Horr's Island in Final Report on Horr's Island: The Archaeology of Archaic and Glades Settlement and Subsistence Patterns." Florida Museum of Natural History, Dept. of Anthropology. Paper, June 1991.
Sassaman, Kenneth E. *Early Pottery in the Southeast and Innovation in Cooking Technology*. Tuscaloosa: University of Alabama Press, 1993.
Savage, Charles. "Dade Residents Recall Submarine Saga with Pride." *The Miami Herald*, July 25, 1999, 4E.
Scarry, C. Margaret, and Lee Newsom. "Archaeobotanical Research in the Calusa Heart-

land." *Culture and Environment in the Domain of the Calusa.* William H. Marquardt, ed. Institute of Archaeology and Paleoenvironmental Studies, Monograph 1, 1992, 375–401.

____, ____, and Marilyn A. Masson. "Calusa and Tequesta Plant Use: Evidence Gleaned from Archaeobotanical Data." Paper presented at the 46th Annual Meeting of the Southeastern Archaeological Conference, 1994.

Schell, Rolfe. *1000 Years on Mound Key.* Fort Meyers Beach, FL: Island Press, 1968.

Schumacher, Carl. *Stories of Life in South Dade.* Barbara Youngbluth and Robert Jensen, eds. Florida City: Florida Pioneer Museum, 1992.

Sears, William H. "Fort Center: An Archaeological Site in the Lake Okeechobee Basin." *Ripley P. Bullen Monographs in Anthropology and History.* Gainesville: University of Florida Press, 1982.

____, and James McGregor. "An Archaeological Survey of Biscayne National Seashore." MS. on file, S.E. Archaeological Center, N.P.S., 1974.

Sellards, E. H. "Human Remains and Associated Fossils From the Pleistocene Of Florida." 8th. Annual Report, *Florida State Geological Survey* (1916), 121–160.

Simpson, C.T. *Out of Doors In Florida.* Miami, FL: E.B. Douglas Co., 1924.

Skow, J. "The Land Where Spring Meets Autumn." *Journal of the New York Botanical Gardens* (1924), 25:53–94.

____. "This Florida Spa Holds a Surprising Lode of Prehistory." *Smithsonian* 17 (1986), 72–83.

Souviron, Richard R., D.D.S. Personal communication. June 1994.

Small, John Kunkle. *From Eden to Sahara, Florida's Tragedy.* Lancaster, PA: Science Press Printing Co., 1929.

Snow, Dean. *The Archaeology of North America.* New York: Viking Press, 1976.

Spencer, Robert F., Jesse D. Jennings, et al. *The Native Americans.* New York: Harper and Row, 1965, 419–420.

Squires, Karl. "Pre-Columbian Man in Southern Florida." *Tequesta* 1:1 (March 1941), 39–46.

Sturtevant, William C. "Creek Into Seminole." *North American Indians in Historical Perspective,* Ed. Eleanor Burke Leacock and Nancy Oestreich Lurie. New York: Random House, 1971, 92–128.

Swanton, John R. "Early History of the Creek Indians and Their Neighbors." 1970. *Smithsonian Institution, Bureau of American Ethnology, Bulletin 73.* New York: Johnson Reprint, 1998.

Tankersley, Kenneth B., Ed. "The Puzzle of the First Americans." *Scientific American Discovering Archaeology* 2:1 (Jan./Feb. 2000), 30–75.

Tattersall, Ian. "Rethinking Human Evolution." *Archaeology,* 52:4 (1999), 22–25.

Tebeau, Charlton W. *A History of Florida.* Coral Gables, FL: University of Miami Press, 1971.

____. *Man in the Everglades.* Coral Gables, FL: University of Miami Press, 1968.

Thornton, Russell. *American Indian Holocaust Survival.* Norman: University of Oklahoma Press, 1987.

True, David O. *Memoir of Do. d'Escalante Fontaneda Respecting Florida,* translated by Buckingham Smith. Historical Association of Southern Florida, 1973.

Tully, Lawrence N. *Flint Blades and Projectile Points of the North American Indian.* Paducah, KY: Collector Books, 1986.

Voegelin, Byron D. *South Florida's Vanished People.* Fort Myers Beach, FL: The Island Press, 1972.

Walker, Karen Jo. "The Mystery of the Pineland Canal." *Calusa News* (Florida Museum of Natural History) (1992), 6:3.

Wanless, Harold. "Inundation of Our Coastlines." *Sea Frontiers* 35:5 (1989), 264–271.
____, and Jeffrey J. Davis. "Carbonate Environments and Sequences of Caicos Platform." *Field Trip Guidebook T347*, American Geophysical Union, 1989, 4–5.
Webb, S. David, ed. *Pleistocene Mammals of Florida*. Gainesville: University Press of Florida, 1974.
____. "Mastodon's Gourd Meals Dash Dispersal Theory." *National Geographic* (1993), 183:4.
Weier, Lisa, Ed. "Update on the Circle." *The Archaeological Society of Southern Florida* 25:6 (Sept. 1999), 2–3.
Weisman, Brent Richards. *Unconquered People: Florida's Seminole and Miccosukee Indians*. Gainesville: University of Florida Press, 1999.
Wheat, Jack. "Florida's Geologic Menagerie." *The Miami Herald*, April 30, 1995, 6–7B.
Wheeler, Ryan J. "Early Florida Decorated Bone Artifacts: Style and Aesthetics from Paleo-Indian Through Archaic." *The Florida Anthropologist* 47 (1994), 47–60.
White, David. *The Archaeology and History of Native Georgia Tribes*. Gainesville: University Press of Florida, 2003.
Widmer, Randolph E. *The Evolution of the Calusa, a Non-agricultural Chiefdom on the Southwest Florida Coast*. Tuscaloosa: University of Alabama Press, 1988.
Willey, Gordon R. *Archaeology of the Florida Gulf Coast*. Gainesville: University Press of Florida, 1998.
Williams Wilma B. "Bridge to the Past: Excavations at the Margate-Blount Site." *The Florida Anthropologist* 36:3–4 (1983), 142–153.
____, and Bert Mowers. "Markham Park Mound No.2, Broward County, Florida." *The Florida Anthropologist* 30:2 (1977), 56–78.
Wing, Elizabeth S. *Paleonutrition*. New York: Academic Press, 1979.
Wolkomir, Richard. "New Finds Could Rewrite the Start of American History." *Smithsonian* 21:12 (March 1991), 130–144.
Yeager, C. G. *Arrowheads and Stone Artifacts*. Boulder, CO: Pruett Publishing Co., 1986.
Zimmer, Carl. "How Old Is It?" *National Geographic* (2001), 78–101.

Edible Botanicals Bibliography

I am indebted to Dr. Lee Newsom, Curator, Center for Archaeological Investigations, Southern Illinois University at Carbondale, for her time and patience in reviewing, correcting, and adding to the edible botanicals data.

Angier, Bradford. *A Field Guide to Edible Wild Plants*. Harrisburg, PA: Stackpole Books, 1980.
Austin, Daniel F. "Historically Important Plants of Southeastern Florida." *Southeastern Archaeology* 35 (1980), 100–104.
Bartram, William. *Travels through North and South Carolina, Georgia, East and West Florida, the Cherokee Country, the Extensive Territories of the Muscogulges or Creek Confederacy, and the Country of the Choctaws*. Philadelphia, 1791; London, 1792; 57, 90.
Brown, Robin C. *Florida's First People: 12,000 Years of Human History*. Sarasota, FL: Pineapple Press, 1994.
Bryant, Vaughn M., Jr., and Stephen A. Hall. "Archaeological Palynology in the United States: A Critique." *American Antiquity* 58:2 (1993), 277–286.

Bullen, Adelaide K. *Florida from Indian Trails to Space Age*, vol. 1. Delray Beach, FL: Southern Publishing Co., 1965.

Gifford, John C. "Five Plants Essential to the Indians and Early Settlers of Florida." *Tequesta* 4 (1944), 36–44.

Griffin, John W. "Archaeological Investigations at the Granada Site." *Excavations at the Granada Site, Archaeology and History of the Granada Site I* by J. Griffin, S. Richardson, M. Pohl, C. McMurray, C. Scarry, S. Fish, E. Wing, L. Loucks, and M. Welch. Florida Bureau of Archaeological Research, 1982.

_____. *The Archaeology of Everglades National Park: A Synthesis*. National Park Service, Tallahassee, Fl., 1988, 296–299.

Hann, John H. "The Use and Processing of Plants by Indians of Spanish Florida." *Southeastern Archaeology* 5 (1986), 91–102.

Kavasch, Barnie. *Native Harvests: Recipes and Botanicals of the American Indian*. New York: Random House, 1979.

Mabberley, D.J. *The Plant-Book*. New York: Cambridge University Press, 1990.

Marquardt, William H. "The Development of Cultural Complexity in Southwest Florida: Elements of a Critique." *Southeastern Archaeology* 5:1 (1986), 63–70.

McGoun, William E. *Prehistoric Peoples of South Florida*. Tuscaloosa: University of Alabama Press, (1993), 53–69.

Milanich, Jerald T. *Archaeology of Precolumbian Florida*. Gainesville: University Press of Florida, 1994, 266–267.

Moerman, Daniel E. *Native American Ethnobotany*. Portland, OR: Timber Press, 1998.

Morton, Julia F. *Wild Plants for Survival in South Florida*, 4th ed. Miami: Fairchild Tropical Garden, 1977.

Newsom, Lee, Ph.D., curator, Center for Archaeological Investigations, Southern Illinois University at Carbondale. Personal communications, 1994–1995.

_____. "Analysis of Botanical Remains from Hontoon Island (8V0202), Florida: 1980–1985 Excavations." *The Florida Anthropologist* 13 40:1 (March 1987), 47–83.

_____. "Archaeobotanical Data from Groves Orange Midden (8VO2601), Volusia County, Florida." *The Florida Anthropologist* 47:4 (1994), 404–417.

_____. "The Paleoethnobotany of the Archaic Mortuary Pond." *Windover: Multidisciplinary Investigations of an Early Archaic Florida Cemetary*, ed. Glen H. Doran. Ripley P Bullen series. Gainesville: University Press of Florida, 2002, 192–210, 306–314.

Newson, Lee A., and Margaret Scary. *Homegardens and Mangrove Swamps: Pineland Archaeobotanical Research*. Unpublished manuscript, 2003.

Niethammer, Carolyn. *American Indian Food and Lore*. New York: Macmillan, 1974.

Perry, I. Mac. *Indian Mounds You Can Visit*. St. Petersburg, FL: Great Outdoors Publishing Co., 1993.

Peterson, L.A. *A Field Guide to Edible Wild Plants*. Boston: Houghton Mifflin, 1977.

Russo, Michael, Ann Cordell, Lee Newsom and Sylvia Scudder. "Glades Ceramics, Archaeobotanical, and Soils Analysis, plus Appendices and References from Horr's Island." *Final Report on Horr's Island: The Archaeology of Archaic and Glades Settlement and Subsistence Patterns*. Florida Museum of Natural History, Dept. of Anthropology, paper, June 1991.

Sleight, Frederick W. "Kunti, A Food Staple of Florida Indians." *The Florida Anthropologist* 6 (1953), 46–52.

Weisman, Brent Richards. *Unconquered People*. Gainesville: University Press of Florida, 1999.

Wunderlin, R.P. *Guide to the Vascular Plants of Central Florida*. Gainesville: University Press of Florida, 1982.

Index

Abaco 154
Accelerator mass spectrometry (AMS) 5, 6
Adams Key 122
Addison's Place 101
Africa 7, 154, 178
African savanna 41, 179
Agatized coal 55
Ais (Ays, Jece) 79, 152, 157, 168, 169, 170, 172
Alabama 86, 166, 172
Alaska 19, 24
Alligator 49, 82
American lion 27, 34, 35, 179
American Philosophical Society 70
Andros Island 40
Anno Domini 6
Antilles 68, 77
Apache 21
Apalachees 172, 176
Appalachian Mountains 15
Arawak 70, 77, 154, 156
Arch Creek 112, 127, 128, 129
Archaeological and Historical Conservancy 3, 31, 32, 37
Archaeological Society of Southern Florida 38
Archaeology 175
Archaeology Day 38
Archaic 4, 35, 40, 50, 51, 52, 53, 55, 56, 57, 61, 66, 79
Archaic Period 50, 52, 64, 72, 73, 83, 97, 98, 102, 118
Archive of the Indies 159
Arctic 19
Argentina 21

Armadillo 37
Artifacts 52
Asia 19, 22, 178
A'teik-ha'ta 25
Atlantic Coastal Ridge 13, 14, 15, 32, 58, 59, 61, 112, 117, 124, 126, 134, 139, 140, 142, 179
Atlantic Ocean 9, 13, 22, 25, 29, 51, 68, 81
Atlantis site 58, 59, 61, 132
Atlatl 33, 39, 40, 44, 50
Aucilla River 21
Audubon, John James 114
Australia 22
Australopithecus 17
Aztec 33

Bahamas 13, 15, 40, 154, 155, 169, 173
Bartram, William 173, 174
Battey's Landing 91
Bay West site 54, 55
Bear Lake Mounds 114, 115
Beaver 41
Belize 120
Belle Glade 89, 127
Bellon, D.H. 151
Beriault, John 79, 141
Bering Strait 19, 68, 178
Beringia 19, 21, 22
Big Circle Mound 84
Big Cypress National Preserve 104, 175
Big Cypress Swamp 13
Big Mound Key 110, 156
Big Pine Key 15, 120
Biscayne Bay 12, 15, 18, 30, 58, 59, 124, 139, 148, 154, 156, 158, 167, 169

219

Biscayne National Park 1, 112, 122, 123, 126
Bison 33, 34, 43
Blanchard, Charles E. 101
Bleeding Tooth 119, 121
Blount, Bruce 81, 82
Bluefish Caves 19
Boca Chica Key 119
Boca Ratones 152
Bonita Springs 68
Boomerang 44, 46
Botanicals, edible 162, 163, 164, 171
Bowlegs, Billy 175
Brazil 21
Breathmaker 176
Brickell Point site 85, 139, 140, 145, 146
Brief History of the Human Race 180
British 166, 171, 174
British Columbia 25
Brooks, H. K., Dr. 38
Broward County 61, 79, 81, 129
Broward County Archaeological Society 83
Broward point 135
Brown, Robin C. 97, 165
Bryozoans 13
Buck Key 94, 110
Buskita 176
Busycon contrarium 64, 129

C14 5, 25, *40*, 74
Calderon, Gabriel Diaz Vara 167
Calendar years before present 5
Caloosahatchee cultural area 57, 58 70, 79, 90, 91, 96, 103
Caloosahatchee River 91, 100, 169
Calos 99, 167
Calusa (Caloosa) 57, 70, 79, 87, 90, 93, 95, 99, 103, 110, 111, 122, 128, 152, 154, 156, 157, 159, 160, 166, 167, 169, 170, 171, 172, 176, 182
Calusa News 57, 93, 101
Camelidae 33
Canada 21
Canoe 64
Cape Coral 100
Cape Florida 54, 149, 166
Cape Haze 79
Cape Sable 79, 112, 113, 114, 115, 122, 171
Captiva 91

Carcharodon megalodon 41
Caribbean 15, 68, 69, 70, 77, 145, 154, 156, 166, 170
Caribbean Monk Seal 137
Caribs 155, 156
Carlos 87, 89, 157
Carlson, Leonard, & Norma 123
Carr, Bob 5, 32, 33, 34, 37, 77, 79, 84, 90, 131
Cash Mound 94, 110
Castaway Mound 113
Cayo Costa 91
Cazambas Mound 101
Celts 63, 64, 106
Central America 65, 68, 69, 142
Ceramic sequence 78
Charleston 166, 168
Charlotte Harbor 41, 79, 91, 100, 103, 110, 156
Chatham River 110, 112
Cheetum site 61
Cherokee 175
Chert 32
Chickasaw 175
Chile 21
Choctaw 175
Chokoloskee Island 101, 110
Christ 6
Cladium jamaicensis 53, 111
Clausen Carl 42
Clewiston 84, 89
Clovis points 22, 23, 24, 25, 29, 31, 74, 78
Coastal Lowlands 12, 13
Colombia 69
Columbus 153, 154, 156
Companion Encyclopedia of Anthropology 2
Composites 44
Conquistadors 156
Continental Congress 174
Cook, Michael 180
Coontie 77, 87, 164, 182
Coprolite 38, 84
Cordell, Ann 94
Cordilleran Ice Sheet 19
Corps of Engineers 81
Court of the Pile Dwellers 104
Creek 166, 172, 176, 182
Cuba 71, 77, 88, 89, 154, 157, 170, 171, 173

Index 221

Cultural materials 37
Cultural periods 4, 5
Curcubita pepo 22, 48, 80, 90, 93
Cushing, Frank Hamilton 69, 70, 91, 93, 98, 100, 103, 104, 105, 109, 154
Cutler 15, 33, 34, 35, 40, 126, 127
Cutler fossil site 30, 31

Dade Circle 84
Dalton points 31
Daughters of the American Revolution 148
Davis, Kim 37
de Allyon, Lucas Vasquez 155
de Aviles, Pedro Menendez 158, 166
deBry, Theodore 181
Deep Lake 49
Deer 33, 44, 50, 64, 82, 93, 137
Deering Estate 30, 126
de Evia, Joseph Antonio 173
de la Puente, Juan Elixio 170
de La Vega, Grascilaso 157
de Luna, Tristan 152
Demere Island 91, 110
Demorey Key 96, 122
de Narvaez, Panfilio 158, 176
Dendrochronology 6
Dentition 40
de Soto, Hernando 152, 176
de Velasco, Lopez 103, 159, 165
Diamondback rattlesnake 43
Diaz, Celestino 141
Dickinson, Jonathan 157, 167
Dinosaurs 19
Dire wolves 2, 27, 34, 35, 37, 179
Dismal Key 113
DNA 52, 53, 180
Dog 22, 33
Douglas, Majorie Stoneman 150
Dugong 38
Dunbar, James 21

East Cape 114
East Okeechobee 79, *81*
Eaton, Jack
Egyptians 6, 96
El Portal mound 148
El Salvador 71
Electron spin resonance 6
Elliot Key 12, 122
Emathla, Eneah 182

English 151, 166, 167, 169, 171, 173
Environmentally Endangered Lands Program 131
Equus 33
Estero Bay 79, 99, 103
Eurasia 17
Everglades 10, 12, 13, 48, 49, 51, 61, 64, 66, 68, 79, 89, 124, 126, 132, 139, 150, 151, 182
Everglades cultural area 111, 112, 125, 129
Everglades Journal 66
Everglades National Park 1, 12, 112, 113, 182
Eyster, Irving 71, 78, 120, 121

Fairbanks, Charles 173
Felis atrox 35
Fiber temper 74, 77
First Seminole War 174
Fisheating Creek 84, 90
Fisher Island 15
Flagler, Henry 1, 133, 139, 147, 178, 182
Flamingo 114
Florida 4, 6, 9, 12, 15, 16, 29, 38, 64, 68, 79, 82, 88, 94, 96, 98, 102, 110, 147, 156, 164, 166, 169, 170, 171, 173, 174, 175, 176, 178, 180
Florida Anthropological Society 3, 180
Florida Archaeological Council 95
Florida Atlantic University 84
Florida Bay 12, 18, 111, 115, 122
Florida Bureau of Archaeological Research 21
Florida City 1, 15, 150
Florida Communities Trust 131
Florida Department of State 105, 134, 136
Florida Geological Survey 12
Florida Horse Conch 125
Florida Keys 1, 117, 118
Florida Museum of Natural History 3, 57, 95, 97, 109, 130, 160
Florida Service 99
Florida State Museum 94
Floridan Aquifer 16
Floridan Peninsula 1, 2, 10, 11, 15, 16, 18, 25, 28, 29, 48, 50, 52, 66, 68, 71, 72, 74, 75, 86, 104, 118, 151, 152, 154, 156, 173, 179
Floridan Plateau 9, 12

Index

Fontaneda, Hernando D'Escalante 87, 88, 122, 131, 156
Ford, James 116
Formative or Ceramic Period 4, 72, 74, 76, 78, 79, 80, 81, 82, 119, 122, 124, 126
Fort Caroline 166
Fort Center 83, 84, 85, 86, 87, 90, 122
Fort Dallas 147
Fort Lauderdale 51
Fort Meyers 90, 96, 99, 101, 157
Fort Pierce 172
Franciscans 159, 160
French 151, 166, 169, 171
French and Indian War 171
Fritchie, John 66

Galt Island 110
Georgia 13, 68, 69, 86, 166, 172, 173
Gifford, John 182
Glades County 84
Glades I, II, and III 4, 76, 77, 78, 82, 122, 135, 137
Goggin, John 119, 121, 126, 130
Goodland Mound 101
Gopher Key 110, 113
Gordon's Pass Canal 101
Gossypium hirsutum 80
Granada site 59, 131, 133, 135, 136, 137, 139, 147, 148, 150, 161
Granberry, Julian 172
Graves Museum of Archaeology and Natural History 3
Green Corn Dance 131
Griffen, John W. 128, 137
Ground sloth 43
Guatemala 71, 120, 141
Gulf Coastal Corridor 28
Gulf of Mexico 25, 28, 51, 54, 65, 68, 89, 91, 93, 100, 104, 110, 112, 113, 115, 142, 154
Gulf Stream 15

Haiduven, Richard G. 59, 60
Hamilton Mound 110
Harper's Weekly 151
Havana 158, 159, 160, 170, 171
Hendry County 84
Hernando point 138
Herrera, Antonio 153
Hickory nut 44

Hispaniola 71, 153, 154, 156, 181
Historic Period 7, 134, 152
Historical Museum of Southern Florida 3, 133
Hitchiti 172, 176
Holmes and Hrdlicka 23
Homestead 1, 13
Homestead Canal 114
Homestead Pioneer Museum 3
Homestead site 115, 116, 117, 124
Hominid 17
Homo erectus 17
Homo habilis 17
Homo neanderthalensis 17
Homo sapiens 17, 19
Honduras 88
Honey Hill site 138, 139, 150, 161, 171
Horr's Island 56, 57, 58, 68, 94, 96, 98
Horse 33, 34, 37
Horse Conch 105, 135, 142
Hudson, Charles 169
Huguenots 166

Iberian Peninsula 22, 178
Ice Age 9, 19, 26, 28, 38, 50, 179
Ilex vomitoria 168
Indian Canal site 101
Indian Creek site 126, 127
Indian Ocean 22
Ingold, Tim 2
Inuit 2

Jackson, Andrew 174
Jaega (Jove, Hoe-bay, Hobe, Xega) 152, 166, 167, 168, 169
Jaguar 34
JC virus 21
Jesuits 156, 158, 159, 160, 165, 171
Job, Herbert 113
Johnson's Key 100, 110
Josslyn Island 91, 94, 96, 98, 110
Jumper, Josie 176
Jupiter 68, 167

Kenworthy, Charles, J. 101
Key Biscayne 15, 54, 127, 139, 149, 150, 161, 166, 171
Key Largo 119, 120, 122
Key Marco 70, 104, 105, 109, 154
Key Vaca 119, 171
Key West 15, 119, 121, 170

Index

Kintigh, Keith 23
Kissimmee River Basin 79, 86
Koreshan State Historic Park 99, 101

Labrador Current 15
La Florida 154, 156, 166, 170, 171, 174, 181
Lagenaria siceraria 48, 80, 90, 93
Lake Okeechobee 10, 18, 48, 49, 50, 51, 53, 66, 75, 79, 83, 84, 86, 89, 90, 100, 103, 104, 111, 169
Lake Sampson 49
Lake Worth 127
Lamme, Vernon 70, 127, 176
Land tortoise 41
Laxon, D.D. 130
Le Moyne, Jacques 181
Levy point 115, 116
Lignumvitae Key 121
Little Climatic Optimum
Little Ice Age 93
Little Manatee River 101
Little Salt Spring 40, 41, 42, 43, 44, 51, 54
Livona pica 121
Llamas 27, 35, 36, 37
Longboat Key 119
Lord, James 102
Lorentide Ice Sheet 19
Lost Lake 49
Lostman's Key 110
Lostman's River 113
Louisiana Territory 171
Lower Creek 169
Lower Matacumbe 120
Lucayas 155
Luer, George 93

Macon, Georgia 144, 145
Macrocallista nimbosa 64, 129, 130
Madden's Hammock 67, 130, 131
Maine 68
Maize 84, 122, 164
Mammoth 23, 26, 28, 30, 33, 41, 43
Man in the Everglades 110
Manatee 38, 41, 165
Manatee County 95
Maples, William R. 97
Marathon 15
Marco Island 56, 69, 100, 101, 105, 106, 112, 122

Margate-Blount site 81, 82, 83
Markham Park site No. 2, 61, 66, 72
Maroons 175
Marquardt, William 93, 110
Martires 156, 167
Martyr, Peter 155
Mastodon 22, 26, 30, 41, 43
Matacumbe 156, 158, 170
Matanzas 166
Matheson Hammock 112
Maya 71, 120
Mayaca 166
Mayaimi 50, 87, 89, 90, 132, 152, 166, 167, 172
Meadowcroft Rock Shelter 21, 22
Melbourne 30, 40
Mendez, Pedro 102, 103
Mesoamerica 22, 71, 72, 154
Mexico 70, 71, 84, 120, 154, 160, 164
Miami 1, 13, 15, 30, 58, 61, 84, 126, 128, 132, 133, 147, 148, 154
Miami Beach 15, 127, 127, 139
Miami Circle 14, 84, 143, 146, 147
Miami-Dade County 5, 12, 13, 32, 61, 71, 84, 126, 130, 131, 146, 147, 150, 165, 166
Miami-Dade County Historic Preservation Division 3, 117, 140
Miami-Dade Public Library 3
Miami Lakes 67, 131
Miami Limestone 13, 15
Miami Museum of Science and Space Transit Planetarium 3
Miami River 58, 61, 85, 126, 132, 133, 134, 139, 140, 147, 160
Miccosukee 152, 172, 176, 182
Middle Cape 114
Mikasuki 172, 177
Milanich, Jerald 160, 175, 177
Mitchell, Scott 130
Monachus tropicalis 137
Mongoloid 35, 40
Monkey Jungle 34, 35, 37, 40
Morton, Dr. Julia 161
Mosquitoes 49, 113, 150, 151
Mound Key 90, 96, 99, 100, 101, 103, 110
MtDNA 180
Mud Lake Canal 115
Muskogean 172
Muskogees 25

Index

Muspa 166
Myakka River 91

Nairne, Thomas 168, 169
Naples 54, 56, 101
Narrative of LeMoyne surnamed DeMorgues 181
National Museum of Natural History 109
National Register of Historic Places 40
Native Americans 56, 152
Native Peoples, Cultures, and Places of the Southeastern United States 177
Navajo 21
Neanderthal Man 17
Nerita peloronta 120
New Mexico 22, 24
New World 19, 33, 166
New York Botanical Gardens 53
Newnan's Lake points 52
Newsom, Lee Ann 92, 164
North America 9, 16, 34, 43, 53, 68, 74
North American Paleoindians 19
North Captiva 91
North Carolina 21
North to the Night 1

Oklahoma 175, 176.
Old Crow Basin 19
Old Rhodes Key 122
Old World 6, 19, 24, 28, 74
Ooids 13, 59
Oolite 13, 15
Orinoco 70, 77
Ortona Indian Mound Park 90
Ortona site 89, 90
Osceola 175
Ossachille 95

Pacific Ocean 68, 178
Paleoindian 4, 19, 25, 29, 32, 35, 40, 41, 43, 48, 49, 50, 56, 72, 73, 126, 172, 178, 179
Paleoindian Period 17, 49, 64, 72, 83, 118
Palynology 80
Pangea 9
Payne, Claudine 57
Peabody Museum of Archaeology and Ethnology 133
Peace Camp site 129, 130
Peace River 91

Peccaries 33, 34, 35, 37
Pellets 13
Pennsylvania Historical and Museum Commission 21
Pensacola 150, 172
Perrine, Henry 126
Perrine Marl 15
Perry, I. Mac 97
Pine 44
Pine Island 69, 90, 91, 92, 95, 96, 98, 99, 100, 103, 110, 159
Pineland site 91, 92, 93, 95, 96, 110
Plantation Key 119, 120, 121
Pleistocene 2, 19, 28, 35, 40, 96
Pleistocene lions 2
Pleuroplaca gigantea 135, 137
point 45
Polar bears 2
Pollen 29, 44, 49, 50
Ponce de Leon, Juan 89, 152, 153, 156, 157, 158, 160, 176, 181
Port Royal 166
Pottery 75, 85
Puerto Rico 153, 154
Pumice 123, 128
Punta Gorda 69
Purdy, Barbara 97

Queen Conch 60, 63, 82, 96, 124, 125, 128, 129, 137, 142

Rabbit 43
Radiocarbon analysis 5
Randell Archaeological Research Center 95
Red mulberry 47
Redland Mound 118
Redlands 117
Renfrew and Bahn 4, 102
Republic of Haiti 153
Ricisak, John 140, 142, 146, 147
Rio Seca 166
River Jordan 88, 89
River Ratones 132
Robertson, William D.D. 152, 153, 165
Rock Lake 49
Romans, Bernard 114, 119, 132, 168, 171, 172
Ronto Developments Marco 56
Rookery Mound 113
Rotary Club of Homestead 1

Index

Royal Palm Hotel 147
Ruskin Thomas Mound 100
Russell Key 113
Russo, Mike 57, 99

Saber-toothed cats 2, 27, 28, 35, 38, 179
Sable Creek 114
St. Augustine 154, 158, 166, 168, 169
St. Petersburg 100
Saltville 21
San Carlos Bay 69
Sand-tempered ware 76, 77, 85, 86, 90, 94, 145
Sandfly Island 113
Sands Key 122, 124
Sanibel 91, 99
Santa Luce 166
Santa Maria site 58, 59, 60, 61, 73, 132
Sarasota County 41, 95
Savannah 169
Sawgrass 53, 111
Scandic climatic episode 86
Sears, William H. 84, 85
Second Seminole War 175
Seminole 129, 130, 131, 148, 152, 165, 172, 173, 174, 175, 176, 177, 182
Seminole Indian War 60
Shark 64
Shark River Slough 53, 111, 112
Short-faced cave bears 35, 37
Siberia 19, 20, 178
Silver Bluff 14, 58, 132
Simon, Alva 1
Simpson, Charles Tory 150
Sloth 27, 34
Small, John Kunkle 53, 98, 114, 115, 126
Smithsonian 105
Snapper Creek site 58, 59, 124, 125, 126, 132
Snow, Dean 153, 161
Soldier Key 15
Solutreans 22, 24
South America 68, 70
South Bayshore Drive 14
South Carolina 29, 68, 69, 168, 169
South Florida 32, 38, 50, 93, 115
The Southeastern Indians 169
Southwest Florida Project 3, 99, 110
Souviron, Richard R., DDS 97

Spaniards 28, 33, 80, 84, 87, 88, 103, 130, 151, 152, 153, 154, 155, 156, 157, 160, 164, 165, 169, 174, 181
Spanish bayonet 80, 114
Spanish Florida 166
Squires, Karl 127
Stalactites 38, 40
Stalagmites 40
Stock Island 121
Strait of Magellan 21
Strobel, Dr. 54, 164
Strombus gigas 120, 129, 137
Sturtevant, William C. 176
Sugarloaf Key 120
Sumerians 6
Summerland Key 120
Swanton, John R. 25

Taino Tribe 156
Tampa 29
Tampa Bay 18, 51, 55, 101, 165, 173
Tancha
Tarpon Lake 49
Taylor Slough 53, 111
Tebeau, Charlton 110
Tectonic 9
Ten Thousand Islands 12, 58, 70, 79, 104, 110, 113, 115, 122
Tequesta 58, 64, 79, 110, 126, 128, 130, 132, 133, 134, 135, 136, 138, 140, 142, 143, 144, 147, 148, 150, 151, 152, 156, 158, 160, 167, 168, 170, 171, 176
Thermoluminescence 6
Third Seminole War 175
Thompson, Edward 148
Thompson, Wiley 175
Throwing stick 47
Timucua 176, 181
Tocobaga 172
Topper 21
Tortoise 43, 47
Totten Key 122
Treaty of Moultry Creek 175
Treaty of Paris 174
Tree ring analysis 6
Turner site 110, 113, 176
Turtle 43

Uchises 170
United States 7, 40, 57
United States National Museum 23

University of Florida 3, 21, 38, 92, 95, 109, 110, 130
University of Florida Institute of Archaeology 110
University of Miami 3, 5, 42
University of New York 15
University of Pennsylvania 69, 109
University of Texas 40, 120
University Press of Florida 177
Upper Matacumbe 119, 121
Useppa Island 91, 94, 96, 97, 98

Venezuela 70
Venice 30, 38, 41
Vero Beach 30, 40
Virginia 21
Virginia Key 15, 126, 139, 149
Vizcayanos 167
Voice of the River

Walrus 2
Warm Mineral Spring 38, 39, 41, 43
Watson place 110
Wax myrtle 44
Webb, David 21

Weisman, Brent Richards 177
Wells, Spencer 22
West Indian Top Shell 120
Weston Pond 51
White-tailed deer 43
Whitewater Bay 115
Whitman 110
Widmer, Randolf, F. 102
Wild cotton 80, 114
Wild Plants for Survival in South Florida 161
Wolf 34
Wood Ibis *43*

Yamassee 169, 170, 176
Yaupon Holly 168
Yellow Fever Creek 100
Yuca aloifolia 80
Yucatan 70, 71
Yuchis 166, 176

Zamia integrifolia 77, 87
Zea mays 164
Zeiller, Todd 129